# HIGHER
# Geography

grade **booster**

✕ Ian Rae ✕

02/120508

ISBN 978-1-84372-480-3

*Published by*
Leckie & Leckie Ltd, 3rd floor, 4 Queen Street, Edinburgh, EH2 1JE
Tel: 0131 220 6831 Fax: 0131 225 9987
enquiries@leckieandleckie.co.uk www.leckieandleckie.co.uk

*Special thanks to*
Euan Kirkpatrick (content review), Pumpkin House (illustration),
BRW (design and page make-up),
Roda Morrison (copy-editing), Noel Keenan (proofreading), Caleb Rutherford (cover design).

A CIP Catalogue record for this book is available from the British Library.

Leckie & Leckie Ltd is a division of Huveaux plc.

Leckie & Leckie has made every effort to trace all copyright holders. If any have been
inadvertently overlooked, we will be pleased to make the necessary arrangements.
We would like to thank the following for permission to reproduce their material:
SQA for permission to reproduce past examination questions
(answers do not emanate from SQA).

# CONTENTS

Introduction ....................................................... 5

1 Atmosphere ..................................................... 15

2 Hydrosphere .................................................... 27

3 Lithosphere .................................................... 37

4 Biosphere....................................................... 45

5 Population ...................................................... 53

6 Rural Geography................................................ 63

7 Industry........................................................ 73

8 Urban Geography ............................................... 85

9 Rural Land Resources .......................................... 97

10 Development and Health ....................................... 111

Conclusion ....................................................... 123

# Introduction

*Why do I need this book?*

*Layout of the book*

*How to use the book*

*The external examination*

*Internal assessment*

*General advice*

## WHY DO I NEED THIS BOOK?

Well, you have completed the course, been inspired by the good lessons, gained energy snoozing through the boring ones and swotted the notes. Only one thing remains – to pass the exam! You should have a multitude of geographical facts jostling round your brain, desperate to spill out on to an exam script. But you need to be able to match what you know to what is being asked in the exam paper and that is where this book will help you. No matter how much you know, you still need to have a good exam technique to demonstrate that knowledge.

Also, as you will be well aware, there is much to learn and this book will help you to prioritise your revision, as well as develop good practice in answering questions. The Higher syllabus has remained much the same since 1991, with only relatively minor changes and the examination questions have retained the same style. Thus it has been possible to analyse the past papers to identify the questions that are most commonly asked. This will enable you to prepare for the exam in a focused way.

## LAYOUT OF THE BOOK

General advice in relation to assessment and exam technique is given in this Introduction.

Chapters 1 to 8 cover each of the topics examined in Paper 1. Chapters 9 and 10 cover the two most frequently tackled topics from Paper 2.

Each chapter:

- identifies the questions most commonly asked and gives an indication of their frequency;
- decodes questions to clarify what you are expected to do and shows how the same question can be worded in different ways;
- gives examples of weak answers;
- gives examples of strong answers;
- gives examples of answers where candidates have 'gone the extra mile';
- provides comment on all the sample answers, highlighting common mistakes, identifying strengths and indicating where improvements could be made.
- Where necessary provides a glossary giving definitions of terms used in the external exam and the sample answers. It is assumed that you will know most of the key terms from your coursework, so the glossary is restricted to terms with which you are less likely to be familiar or which are particularly crucial.

## HOW TO USE THE BOOK

This book is not a coursework guide. It focuses entirely on the exam itself and should be used in conjunction with your course notes. Obviously it can be used at the end of the course to polish your exam technique, but it would also be very helpful if used throughout the course. Once you have completed a topic in school, you could refer to the relevant chapter. Attempt to do the questions yourself and only then look at the sample answers and read the comments. These will mean much more to you if you have first grappled with the question yourself. In this way you will be developing your answering skills throughout the course and this will stand you in good stead for your preliminary examination and NABs, as well as the final external examination.

# THE EXTERNAL EXAMINATION

There are two papers. Paper 1 lasts 1 hour 30 minutes and covers eight topics divided between Physical Environments and Human Environments. Paper 2, entitled Environmental Interactions, lasts 1 hour 15 minutes and includes six topics divided between Physical Interactions and Human Interactions.

| PAPER 1 | | PAPER 2 | |
|---|---|---|---|
| PHYSICAL ENVIRONMENTS | HUMAN ENVIRONMENTS | PHYSICAL INTERACTIONS | HUMAN INTERACTIONS |
| Atmosphere | Population | Rural Land Resources | Urban Change and its Management |
| Hydrosphere | Rural Geography | Rural Land Degradation | European Regional Inequalities |
| Lithosphere | Industry | River Basin Management | Development and Health |
| Biosphere | Urban Geography | | |

## Paper 1

This is in three sections. Six questions in all are to be attempted, totalling 100 marks.

Section A contains two questions on Physical Environments and two questions on Human Environments. **All four** of these questions must be attempted. There are usually between 16 and 20 marks for each question, which may be in one or two parts.

Section B contains two questions on the two Physical Environments topics not tested in Section A. Only **one** of these should be attempted. There are usually 14 marks for each question, which may be in one or two parts.

Section C contains two questions on the two Human Environments topics not tested in Section B. Only **one** of these questions should be attempted. There are usually 14 marks for each question, which may be in one or two parts.

The topics that occur in each section will vary from year to year. **Therefore you cannot ignore any topics when revising.**

This paper always contains at least one question based on an **Ordnance Survey map** in either the Hydrosphere, Lithosphere, Industrial or Urban questions. It is usually scale 1:50 000.

## Paper 2

This is in two sections. **Two** questions are to be attempted, one from each section, totalling 100 marks.

Section 1 contains three questions covering each of the Physical Interactions, from which you answer one, which may be divided into three or four parts, totalling 50 marks.

Section 2 contains three questions covering each of the Human Interactions, from which you answer one, which may be divided into three or four parts, totalling 50 marks.

**2008 is the first year that this marking system will apply. In previous years half a mark was awarded for each valid point in an answer, but from now on it will be a whole mark, although nothing more is being demanded of you. Consequently the total marks are double what they have been in the past, so bear this in mind when you are looking at past papers. Past questions used in this book have been amended to the new marking system.**

# INTERNAL ASSESSMENT

During your coursework you will have **three** internal assessments called National Assessment Bank Items (NABs) which are written by the Scottish Qualifications Authority (SQA) and marked by your teachers. There is one on Physical Environments, one on Human Environments and one on Environmental Interactions. You only have to achieve 50% in **each** of these and there is one opportunity to be reassessed in each, but do **take them seriously.**

First, you **must** pass them all. No matter how well you do in the external examination, if you have not passed all the internal assessments you will not gain a Higher award.

Secondly, they are a good opportunity for you to demonstrate both to yourself and to your teacher that you can cope with the jump in standard from Standard Grade to Higher.

Thirdly, and very importantly, they are good practice for your preliminary exam and the final external exam itself, enabling you to practise pacing yourself against the clock.

# GENERAL ADVICE

## Watch the clock

**Time** is the biggest enemy you face in the exam, especially for able candidates who are capable of writing extended answers to all the questions. It is absolutely imperative to make sure you **answer all the required questions**. Failure to do this can have a big impact on your final grade. If you have written nothing in response to a question you cannot receive any marks (Answer 55).

You only have a limited time available so keep your eye on the clock and pace yourself carefully. In Paper 1 you must answer six questions in 1 hour 30 minutes. That is only 15 minutes for each question and it may well contain two parts. In Paper 2 you have to answer two questions in 1 hour 15 minutes and each of these will certainly contain three parts or, more probably, four – that is 37 minutes per question and 9 minutes per part on average!

Before you start each question consult the clock and impose a time-limit on yourself, depending on the number of marks in the question. If you can still write more when your allocated time is up **leave space** at the end of your answer so that you can return to it later if you have time.

Now that you appreciate its importance, **don't waste time**.

- **Choose your questions quickly.** In Paper 1 you have a choice in Sections B and C. There may be much to read in these questions, which can eat up valuable time. If you have revised thoroughly you should be capable of tackling any of the questions, so don't agonise over the decision. Make your choice quickly and get on with it. In Paper 2 you probably will not have a choice, unless you have studied more than two Interactions at school, in which case you should not take too long considering the alternatives.

- Be clear **before** the exam about what the **instructions** are likely to be, e.g. how many questions you should answer in each section.

- **Don't repeat yourself.** You will not get any more marks for saying the same thing over and over again. This applies also to annotated diagrams. If you have produced a fully annotated diagram you will not get any more marks for repeating this information in the written part of your answer. (Answer 15 shows how annotated diagrams can obtain full marks.)

- **Don't rewrite the question** (Answers 17 and 42).

- Make sure your answer is **relevant** (Answers 26 and 29).

- If you get a **mental block** don't rack your brains indefinitely – **leave space** and come back to the question later.

## Read the question carefully

Teachers have been giving this advice since their pupils were scratching the answers on slates and it is still vitally important. If your answer is not responding to the question asked you will get no marks. You should identify the **key words** in the question. Take the question below.

> *International* migrations may be voluntary or forced.
>
> Referring to one named *example* of *each* type of migration, *explain* why the migration took place.

The key words have been italicised. Failure to spot the significance of any of these words would mean that the answer would not be tackled properly and marks lost. Note that some key words are highlighted in the exam paper, but not necessarily all of them – you will have to spot most of them yourself (Answers 16, 27, 29, 31, 42, 45, 47 and 51).

If the question is in several parts, make sure you have answered each one, otherwise again marks will be forfeited (Answer 36).

Do not assume that each part of a question will be on the same topic. In 2004, part (a) of the Atmosphere question related to circulation cells and part (b) to ocean currents and was accompanied by a reference diagram of ocean currents. Distracted by the diagram, many pupils wrote about ocean currents in both parts.

## Give detailed answers

There is a big jump in standard from Standard Grade to Higher, so that answers require much more detail at Higher level (Answers 7 and 12). In particular, it is necessary to provide **examples** of specific case studies. Even when not asked to provide them it is good practice to do so since it creates an impression of a knowledgeable candidate and may sway the marker to award marks in marginal situations (Answers 3, 30, 43, 44, 48, 49 and 54).

## Give proper explanations

A common failure is to give descriptions rather than explanations. If asked to explain something you should be giving reasons. Consequently, you should be using linking words and phrases such as 'because', 'this means that', 'therefore', 'so', 'so that', 'due to', 'since', 'the reason is'. If you check back over an answer to an 'explain' question and you realise that you have not used any of these words then you probably have not answered the question (Answers 10, 25, 32 and 41).

## Use the resources provided

Many questions will ask you to 'describe and explain' so don't forget to describe as well. Some questions, like at Standard Grade, will be based on resources such as maps, diagrams or tables so there can be some easy marks available to be lifted from this data. Don't pass these up (Answers 2, 4, 9, 23, 24, 35 and 50).

## Develop points

You should write proper sentences developing the point you are trying to make. **Do not write 'lists'.** You will be penalised for this (Answers 26, 34 and 45).

## Draw diagrams

Sometimes you are specifically instructed to use annotated diagrams in your answer, so you will be penalised for not doing so (Answers 1, 14, 15, 18 and 45). However, diagrams may also be appropriate in other answers and you will be given credit for providing them (Answer 33). Practise drawing relevant diagrams, keeping them simple and easy to reproduce quickly in the exam.

NB: An annotated diagram is one on which labels and notes are written.

# Plan an answer

Do not take this instruction too literally. You are not expected to produce an elegantly crafted essay and you certainly do not have time to produce one. However it is a good idea, especially in the bigger questions, to quickly **jot down key words** or reminders of the important things to include in your answer. Cross them off as you deal with them and check them before you move on to your next question, so that you have not forgotten anything. For example, if you are asked to describe and explain the climate of a location in West Africa based on a climate graph and map, you could jot down T and RF (abbreviations of temperature and rainfall to remind you to describe both parts of the graph) and ITCZ, cT, and mT (to make sure you explain the influences of the Inter-Tropical Convergence Zone and Tropical Maritime and Tropical Continental air masses).

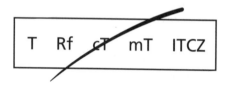

Box these jottings off and put a line through when you have finished so that the marker does not think it is part of your answer.

## Give grid references

It is vital to give these when providing OS map evidence and good answers may not get full marks if they are not given (Answers 5 and 6).

## Realise that marks may be allocated flexibly

For instance, you might have a three-part question that has a total of 24 marks but no indication of how many marks there are for each part. There is unlikely to be a fixed 8 marks for each part. There may be up to 14 marks available for any one part. So, if you cannot answer well on one part, answer fully in the other two parts and you should still finish with a strong mark out of 24 (Answers 11 and 56).

## Select questions that you can do

This may seem blindingly obvious but candidates can do funny things if they panic under pressure in the exam. For instance in Paper 2, a pupil who has studied Rural Land Resources in school may open his exam paper and be thrown perhaps by an unfamiliar reference diagram. In panic he thumbs through the paper and a reference to equatorial rainforests catches his eye in the Rural Land Resources question. 'Aha' he thinks 'I did this at S Grade. I'll do this instead'. He and you should resist that temptation. It is highly unlikely that you will be able to provide sufficient detail to answer a 50 mark question that you have not studied at Higher level. Look again at the question for which your teacher has prepared you and you should find, on careful reading, that you can tackle it.

## Use all the time available

Some candidates seem to think it is 'cool' to be seen leaving the exam early. On the contrary, it is just plain stupid. If you have finished very early the chances are that you have missed out a question or part of a question or have simply not written enough. So, if you do have time in hand at the end of the exam, use it to look back over your answers **and** the questions to make sure first that you have answered all the required questions and their component parts.

Then check that your answers are relevant, that what you have written under pressure actually makes sense, that any illegible writing is corrected and add in any additional diagrams or points that you realise you have omitted.

## Prepare thoroughly

Exam technique will count for nothing if you have not done this.

Many pupils find it difficult to adjust at first to the demands of Higher Grade. At Standard Grade, with its emphasis on Enquiry Skills, most questions are resource based. This means that pupils can do well at Standard Grade by applying the skills they have developed in their coursework. At Higher, however, the emphasis is predominantly on Knowledge and Understanding, requiring the learning and retention of geographical information. In short, you have to **revise** thoroughly or 'swot it up', so that you can provide the necessary detail in your answers referred to earlier. However, the syllabus makes it quite clear what you have to know and be able to do, so that, although there is much to learn, there is a certain predictability about the exam.

You may be wondering how accurately you can predict questions, hoping that you can miss topics out of your revision. Any area of the syllabus may be examined so you should certainly **revise everything** you have done in your coursework. Nevertheless, it is true that certain areas of the syllabus are examined more frequently than others. This book will help you to identify those areas and enable you to **prioritise** your revision.

Always revise with a **pencil and paper** to hand. If your method of revision is just to read over your notes not much is likely to stick. So instead, after reading through a topic try to jot down the key points and then refer back to your notes again to see how much you remembered and what you missed.

Start revising **well in advance** of the exam date so it is not all left to the last minute. Set yourself a timetable for when to revise each topic. This should enable you to get a good night's **sleep** beforehand, leaving you sharp and focused for the exam itself.

# 1 Atmosphere

1 Solar radiation variations

  Exam example 1

2 Global climate change

3 Redistribution of heat energy by atmospheric circulation or ocean currents

  Exam example 2

  Exam example 3

4 West African climate/air masses

  Exam example 4

  Other possible questions

  Exam example 5

Glossary

These questions contain more technical terms than any other part of the exam so, unless you have thoroughly familiarised yourself with the way questions are phrased, they can appear quite daunting. However, once you have grasped what is required the answers can be logical and straightforward.

The reference diagrams and graphs may use units that you have not met before, but do not be phased – all you need to understand is what they are measuring, which should be obvious from the heading on the diagram. Also, pay close attention to diagram keys.

There are four questions that are commonly asked and a couple of others that have occurred two or three times.

# 1 SOLAR RADIATION VARIATIONS

This question may be accompanied by a reference diagram such as A or B below.

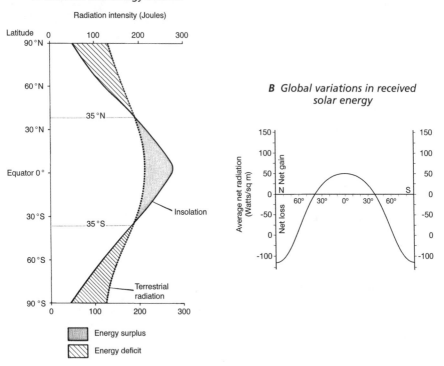

**A** *Latitude and energy balance*

**B** *Global variations in received solar energy*

Both these diagrams, though apparently different, are effectively showing the same thing.

A has lines for both terrestrial radiation and insolation (solar radiation) with the difference shaded to show surplus or deficit. In B the terrestrial radiation has been subtracted from the insolation to produce just one line on the graph showing either surplus (net gain) or deficit (net loss). **Thus both graphs demonstrate that there is a surplus (or net gain) of solar energy in the Tropics and a deficit (or net loss) towards the Poles** and that is what the question will ask you to explain. Other ways this has been asked are:

> **Describe and explain the latitudinal variation of the earth's energy balance.**

or more simply as in the example below.

# EXAM EXAMPLE 1

> With the aid of an annotated diagram, explain why Tropical
> latitudes receive more of the sun's energy than Polar regions.
>
> 8 marks

**Answer 1**   This is a passable answer.

Tropical latitudes receive more sun than the poles. It is always overhead,
unlike the poles where it shines at a low angle (✔). Consequently it passes
through more of the earth's atmosphere at the poles (✔), meaning that
more of the sun's energy is absorbed and reflected (✔). Also at the poles
the albedo effect means that the white colours of the snow and ice
reflect the sun's rays (✔), unlike at the Equator where the dark colours of
the rain forest absorb heat (✔).                    Total: 5 out of 8.

**Why is this only a passable answer?**

From the third sentence onward the answer is well written, but the first
sentence is too vague and the second sentence only receives one mark because
it is not quite correct – the sun is not 'always overhead'. The mid-day sun is
always high in the sky, but is in fact only directly overhead on two days of the
year. Thus it would be more correct to say 'In the Tropics the mid-day sun is
always high in the sky, so that the sun's rays are always very intense and
concentrated in a smaller area than in Polar regions where the sun shines at a
low angle.' This point could have been clearly illustrated with a simple
annotated diagram and the fact that one was not provided, despite being
demanded in the question, means that the candidate could only have gained a
maximum of 6 marks, no matter how good the written answer.

# 2 GLOBAL CLIMATE CHANGE

This is an easy question if you have swotted up your notes on global warming.
You are likely to be given a reference diagram such as the one shown on the
following page.

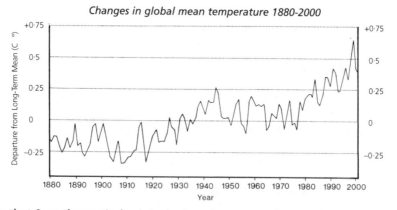

Changes in global mean temperature 1880-2000

Note that 0 on the vertical axis is the long-term mean (average) temperature over the duration of the period shown on the horizontal axis. Also note that the units on the vertical axis are deviations from the mean and **not** actual temperatures.

You may be asked to **describe the variations in global temperature**, in which case you should emphasise the overall increase in T while also noting the fluctuations and quote some figures from the graph to exemplify these.

Also, you would be asked to explain **the human and/or physical factors** which may have led to these changes. So make sure that you can give details about the greenhouse effect and why man is increasing the number of greenhouse gases ($CO_2$, CFCs, methane, and nitrous oxides) in the atmosphere as well as physical factors such as sunspot activity, Milankovitch Cycles and volcanic eruptions. **Do not merely list** the factors, **explain** them.

# 3 REDISTRIBUTION OF HEAT ENERGY BY ATMOSPHERIC CIRCULATION OR OCEAN CURRENTS

Again, these questions can sometimes sound much harder than they really are. For example:

> **Explain how ocean currents operate to maintain the energy balance.**
>
> or
>
> **Describe the role of atmospheric circulation in the redistribution of energy over the globe.**

However, all you are really being required to do is **explain the pattern of ocean currents or surface winds/circulation cells** while emphasising that currents or winds moving away from the Equator move heat energy to higher latitudes and those moving towards the Equator will have a cooling effect on the areas they move towards.

# EXAM EXAMPLE 2

**Study Reference Diagram.**

**Explain how the circulation cells A, B and C and the related surface winds assist in the distribution of energy over the Earth.**                                                                8 marks

*Reference diagram. Atmospheric circulation and surface winds*

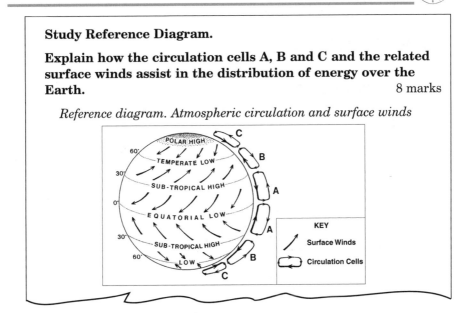

**Answer 2**   This is a weak answer.

Air rises at the Equator and descends back to earth at the Sub-tropical High (✔). Some air returns as surface winds to the Equator, while the rest goes towards the Temperate Low where it rises and circulates in cell B (✔). At the poles air sinks, moves towards the equator and rises at the Temperate Low (✔).                                    Total: 2 out of 8.

**Why is this a weak answer?**

This answer contains no explanation whatever. It is merely a straight description of the diagram, which is why the three descriptive points made have not all been credited and only two marks have been awarded. The cells are not named; the winds are not named; there is no reference to convection to explain the rising air at the Equator nor the density of cold air causing air to sink at the Poles and Sub-tropical High and, most significantly, no indication of how the movements of warm and cold air distribute heat energy over the earth as the question specifically asks.

Nevertheless, you should note that this candidate, despite apparently not understanding anything about the topic, has still managed to salvage a couple of marks by using the resource provided. He has not just thrown his hands in the air and written nothing.

# EXAM EXAMPLE 3

**Study Reference Map.**

**Describe and explain the world pattern of ocean currents shown.**
8 marks

*Reference map. World ocean currents*

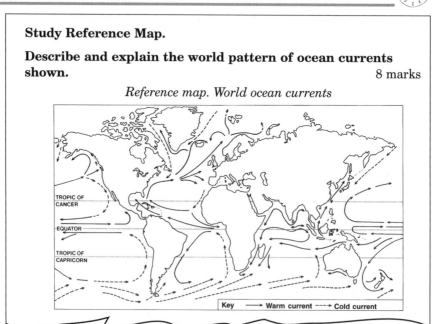

TROPIC OF CANCER

EQUATOR

TROPIC OF CAPRICORN

Key  ——→ Warm current  ----→ Cold current

**Answer 3**   **This is a very good answer.**

At the Equator the Trade winds (✔) push surface water westwards (✔) until they are obstructed by continental landmasses which deflect the currents northwards and southwards (✔). The westerly winds then push the currents eastwards (✔) e.g. in the N Atlantic the SW winds push the Gulf Stream NE towards Europe as the North Atlantic Drift (✔). Currents like these, moving away from the Equator, carry warm water into cooler latitudes (✔). On the E side of the N Atlantic the Canaries Current flows back towards the Equator (✔). Thus there is a clockwise movement of the currents in the oceans of the N hemisphere and an anti-clockwise movement in the S Hemisphere (✔). The Coriolis Force also deflects the currents (✔). Cold water is denser than warm water so cold water sinks away from the Poles (✔) and spreads towards the Equator forming cold currents (✔) such as the Labrador Current (✔).          Total: 8 out of 8

**Why is this a very good answer?**

This answer covers most of the relevant factors and easily achieves full marks. There would probably be a maximum of only 2 or 3 marks available for naming currents so the candidate has not wasted time by naming too many examples.

# 4  WEST AFRICAN CLIMATE/AIR MASSES

For this topic you may be asked to **describe and explain the climate of a particular place in West Africa or explain rainfall variations/distribution**. In either case there will be helpful maps and/or climate graphs provided to help you. In answering either of these questions you will need to refer to the movement of the ITCZ and the influence of the mT and cT air masses.

> Remember, when describing a climate graph, the line is temperature and the bars are rainfall. Also remember to describe seasonal trends and quote figures such as maximum temperature, minimum temperature and annual rainfall.

# EXAM EXAMPLE 4

> Study Reference Maps and Reference Diagram.
>
> With the aid of the climate graph and the maps provided, describe and account for the climate of Jos in Central Nigeria.                                  10 marks
>
> *Continued on next page*

Reference Maps (Selected air masses and fronts over Africa in January and July)

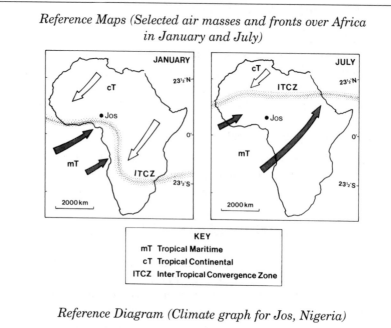

KEY
mT Tropical Maritime
cT Tropical Continental
ITCZ Inter Tropical Convergence Zone

Reference Diagram (Climate graph for Jos, Nigeria)

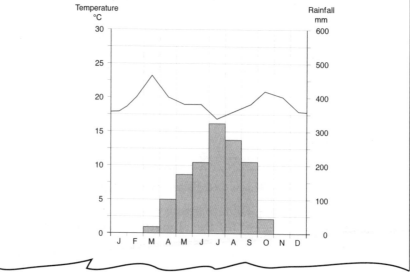

**Answer 4** **This is a very good answer.**

Jos has 2 peaks of T (✔), 23C in March and 21C in October (✔). The min T is 17C in July (✔). There is a drought from November to February (✔) and a wet season with a peak of rainfall in July (✔). The total annual rainfall is approximately 1170 mm (✔).

The dry season occurs while the ITCZ lies to the south of Jos (✔) and the town is affected by the hot dusty Harmattan wind (✔) which blows from the cT airmass (✔). As the position of the overhead sun moves N the ITCZ is dragged N as well (✔), bringing convectional rainfall (✔). As it passes further N warm, moist air (✔) is drawn in from the mT airmass bringing further rain (✔). The lower summer T's are caused by the cloud cover associated with the rain (✔).

Total: 10 out of 10

**Why is this a very good answer?**

There might be as many as 4 marks available for description, so the first paragraph comfortably achieves this. The candidate has accounted for the climate in the second paragraph, so he has clearly separated description and explanation. However, it is not essential to do this. Sentences containing both description and explanation are perfectly acceptable, e.g. 'The dry season extends from Nov to Feb (1), because, at this time, it is influenced by the dry cT airmass (1).'

# OTHER POSSIBLE QUESTIONS

On two occasions candidates have merely been asked to **describe and explain the characteristics of the mT and cT air masses**. This is easy but do remember to include details of actual temperatures and relative humidities.

Another solar radiation question which has occurred twice is

> **Describe and explain the energy exchanges that result in the earth's surface receiving only 50% of the solar energy which reaches the outer atmosphere.**

In 2006 this same question appeared in a less complex form.

# EXAM EXAMPLE 5

**Study Reference Diagram.**

**Explain why the Earth's surface can absorb only 50% of the solar energy received at the outer atmosphere.**   8 marks

Reference Diagram (Earth–atmosphere energy exchanges)

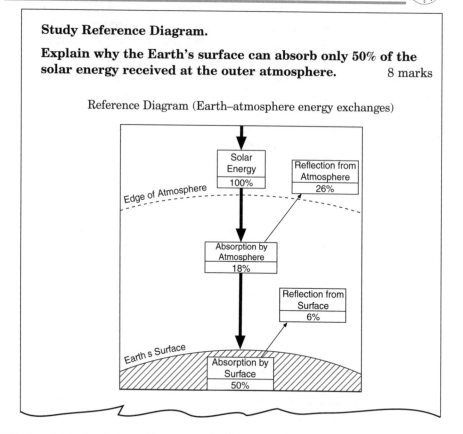

This is a fairly simple question to answer. The diagram is very useful, with probably 4 or 5 marks available for 'straight lifts' from it. The remaining marks would come from simple explanations of the albedo effect and the role of clouds, gases and dust in both reflection and absorption.

# Glossary

**Circulation cell** – A part of the atmosphere within which air circulates vertically, i.e. the Hadley, Ferrel and Polar cells.

**Departure from long-term mean** – How much hotter or cooler the temperature is compared to the average temperature over a long period of time.

**Energy budget/energy balance** – Two terms which relate to the balance between solar and terrestrial radiation.

**Global mean temperature** – The air temperature averaged out over the whole of the earth.

**Isohyet** – A line on a map which joins places of equal rainfall.

**Joule** – A unit of energy.

**Received solar energy** – The solar radiation which passes through the atmosphere to reach the earth's surface.

**Solar radiation/insolation** – Two terms for the short-wave heat energy reaching earth from the sun.

**Terrestrial radiation** – The long-wave heat energy which escapes from the earth's surface.

# 2

# Hydrosphere

1 Ordnance Survey map questions

Exam example 1

2 Explanations of how physical features are formed

Exam example 2

3 Hydrological cycle

4 River flow data

Exam example 3

Glossary

The most commonly asked questions, every two years on average, are those relating to **OS maps** and those which demand explanations of the way **river features** are formed. Every three or four years there are questions about the **Hydrological Cycle** and **River flow data**.

## 1 ORDNANCE SURVEY MAP QUESTIONS

In most instances these involve describing the **physical features** of a river and its valley. Make sure you confine yourself to physical features – do not mention land uses, bridges, settlements, etc. Also, remember to give **grid references**. The table on the following page indicates what to note in your answer.

### Describing rivers on OS maps

**PHYSICAL FEATURES OF RIVER**

| | |
|---|---|
| **Course** | straight or meandering?<br>does it get wider as it flows downstream?<br>direction of flow?<br>speed? flowing down a steep or gentle slope? |

**Features to note if present**

braiding, embankments, oxbow lake, meanders, waterfalls

| | |
|---|---|
| **Tributaries** | how many?<br>from which direction do they come? |

**PHYSICAL FEATURES OF VALLEY**

| | |
|---|---|
| **Trend** | e.g. 'North to South', 'East to West' |
| **Shape** | e.g. U-shaped, V-shaped, gorge<br>steepness of valley sides |
| **Width** | how wide is the valley floor? |

**Features to note if present**

flood plain, interlocking spurs

| | |
|---|---|
| **Height** | valley floor – e.g. 'It descends from 300 m to 290 m'<br>valley sides – e.g. 'They rise from 100 m to over 400 m' |

# EXAM EXAMPLE 1

**Look at OS type map on next page.**

**Using grid references, describe the physical characteristics of the River Booster and its valley from 596020 to573060**

8 marks

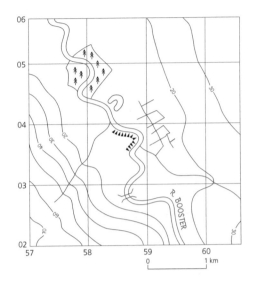

---

**Answer 5** **This is a weak answer.**

The river starts off straight, flowing N and then swings to the W, before going NE and then NW, NE again and then NW, N, NW and back NE and N (✔). The river has a lot of meanders (✔). It flows under a bridge at 587028 and goes through woodland at 580050.          Total: 2 out of 8

**Why is this a weak answer?**

The first sentence demonstrates a common fault – the 'travelogue' description. The 2 marks awarded could have been obtained by simply writing 'The river meanders to the NW.' The references to the bridge and woodland are irrelevant. No grid references have been given to identify meanders and there is no mention of the valley or a number of other features. A poor answer.

**Answer 6** **Here is a much better answer.**

The river meanders (✔) towards the NW (✔). An example of a meander is at 587028 (✔). The river is roughly the same width throughout this section (✔) and could be in its middle course (✔). There are 2 tributaries (✔) with confluences at 588036 and 584040 (✔). There is an ox-bow lake at 585044 (✔). There is an embankment at 587038 suggesting that the river is prone to flooding here (✔). The valley has a flood plain (✔) which widens from 500 m wide in the S to 2 km wide in the N (✔). The E side of the valley is gently sloping (✔) to over 30 m (✔) whereas the W side is steep (✔), rising to over 70 m (✔).

Total: 8 out of 8

**Why is this a much better answer?**

This excellent answer actually makes 15 points worthy of marks, demonstrating just how easy it is to gain marks in a question like this. Note also that, as required by the question, both the river and the valley are referred to and grid references are given for each river feature identified.

The following questions have also been asked:

> **Compare and contrast the physical features of two rivers and valleys shown on the map.**

Essentially you are answering in the same way as above, but for two rivers instead of one. You should point out both similarities and differences, e.g. 'Both rivers flow E in wide meanders, but River A is wider than River B and has a bigger floodplain', and so on.

> **Annotate a base map to show the physical features of a river and its valley.**

This can throw pupils, but is very straightforward. Below is an example of an annotated base map for the River Booster shown on the previous page.

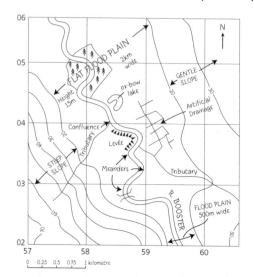

> **Describe evidence that a river is in its lower course.**

This might also be asked about a river's middle or upper course. So, make sure you know the typical characteristics of a river and its valley in each of its three courses and be able to spot these on an OS map.

## 2 EXPLANATIONS OF HOW PHYSICAL FEATURES ARE FORMED

> **Important tips:**
> - Always use **annotated diagrams** in your answer, even when not specifically asked for them.
> - **Practise** drawing the diagrams when you are revising.
> - Use **simple** line drawings – they are much quicker and easier to draw than artistic block diagrams.
> - Give plenty of **detail** on the processes – do not be fooled into thinking 'This is easy, I got full marks for this at S Grade'. There may be twice as many marks for a question like this at Higher.
> - When given a choice of features, choose the one you can write most about.

Make sure you can explain the following features: meander, ox-bow lake, waterfall, levee, river terrace, braiding, flood plain. The first four are those most likely to be specifically asked and often there is a choice.

## EXAM EXAMPLE 2

> **Explain, with the aid of a diagram or diagrams, how a meander is formed.** 6 marks

**Answer 7**   **This is a weak answer.**

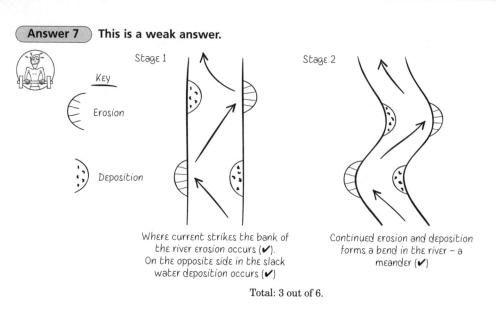

Stage 1

Key

Erosion

Deposition

Where current strikes the bank of
the river erosion occurs (✔).
On the opposite side in the slack
water deposition occurs (✔)

Stage 2

Continued erosion and deposition
forms a bend in the river – a
meander (✔)

Total: 3 out of 6.

**Why is this a weak answer?**

At Standard Grade this would have been a satisfactory answer, but it lacks the detail required at Higher. The changing direction of the current should have been explained by reference to riffles and pools. The processes of river erosion such as hydraulic action could have been mentioned, as could the migration of meanders downstream. A cross-section of the meander could have been drawn to show the varying depth of water across the meander, helicoidal flow and the positions of river cliff and point bar.

# 3 HYDROLOGICAL CYCLE

Usually you are asked to describe, with the aid of a diagram, the global hydrological cycle. You should have notes on this, so swot them up. Make sure you can draw the diagram, annotate it and explain the roles of the sun's heat, icecap storage, evaporation from oceans and lakes, transpiration, cooling causing condensation into clouds and precipitation, interception, surface run-off, infiltration and groundwater.

If asked to **'explain how a balance is maintained within the Hydrological Cycle'**, just answer in much the same way as above, but emphasise that there is a fixed amount of water in the system, so that any change in one part of the cycle will be compensated by change elsewhere, e.g. increased storage in the icecaps would result in a drop in sea level.

In 2006 this question was asked:

> **Describe the movement of water within a drainage basin.**

Again the answer will be similar, but also use the terms inputs, outputs, transfers and storage (see glossary).

## 4 RIVER FLOW DATA

Usually you will be provided with a bar graph of rainfall and **a hydrograph** which is a line graph of a river's discharge over a period of time, either a year or the days immediately after a storm – a storm hydrograph. Alternatively it might be a line graph of changing river level (depth). The question will be to describe and explain the relationship between the rainfall graph and the hydrograph.

## EXAM EXAMPLE 3

Study Reference Diagram.

Describe and account for the changes in discharge levels of the River Wyre between the 18th and the 19th of December 1993.

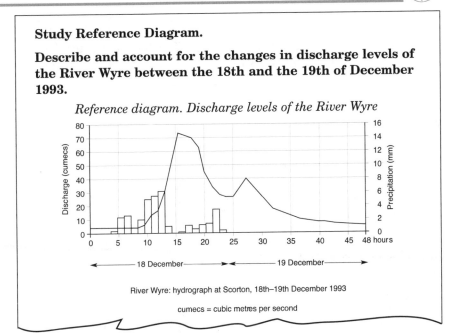

*Reference diagram. Discharge levels of the River Wyre*

River Wyre: hydrograph at Scorton, 18th–19th December 1993

cumecs = cubic metres per second

**Answer 8**    **This is a weak answer.**

*The discharge is zero until 4 Dec when it starts to go up. Peak discharge is 31 cumecs on 12 Dec. There is 15mm rain on 16 Dec which explains why the discharge increases to 18 cumecs on 22 Dec.*    Total: 0 out of 8

### Why is this a weak answer?

This is a disastrous answer! The candidate has completely misread the graph, thinking that the bar graph is the discharge and the line graph the rainfall. She has also thought that the horizontal axis shows dates in Dec – 48 days in December! So, be careful when interpreting combined graphs such as these – look carefully at them before plunging into your answer.

> **Remember** – rainfall totals are always shown as a bar graph.

**Answer 9**    **This is a passable answer.**

*At first the discharge is 4 cumecs (✔), but at 0900 18 Dec it starts to rise (✔), reaching a peak of 73 cumecs at 1600 18 Dec (✔). Thereafter it declines to 27 cumecs at 0100 19 Dec (✔), before increasing again to a second peak of 40 cumecs at 0400 19 Dec (✔). Thereafter it declines to 5 cumecs by end of 19 Dec (✔).*

*The increases in discharge are caused by increases in the rainfall (✔).*
                                                                    Total: 5 out of 8

### Why is this only a passable answer?

Paragraph 1 easily gets the 4 marks available for description, but the only explanation is in paragraph 2 and is very brief so only receives one mark.

The following excellent answer accounts for the discharge changes as it describes them, which is probably the easiest way to tackle this sort of question. Descriptive points and explanatory points have been identified by the letters D and E. Notice also that this candidate, unlike the previous one, has not bothered to work out the times and dates, but just read off the hours – this is quite acceptable.

**Answer 10**

*At first there is no rain so the river remains steady (E) at 4 cumecs (D). The discharge starts to rise after 9 hours (D) due to the rain which started to fall after 4 hours finally reaching the river (E). At first the rain*

would have infiltrated, so would not have raised the river level (E). Rain continued to fall until 14 hours, getting progressively heavier so that the ground would become saturated (E) and run-off would increase rapidly (E), causing the discharge to rise rapidly as well (D) to a peak of 73 cumecs after 16 hours (D). This is 3 hours after the peak rainfall due to basin lag (E). Rain continues, but much less heavily than before, so that discharge drops sharply (E) to 27 cumecs after 25 hours (D). However, another rainfall peak after 23 hours results in a second peak of discharge (E) 3 hours later at 40 cumecs (D) After 24 hours the rain has stopped but the discharge only gradually decreases as groundwater and tributaries continue to feed water into the river (E) until, after 48 hours, it is close to its original level at 6 cumecs (D). **Total: 8 out of 8**

On one occasion **maps** of two different drainage basins and their storm hydrographs were provided and the question was to explain these. So, make sure you know how factors such as tree cover, gradient, rock permeability, drainage density and urbanisation affect the shape of a storm hydrograph.

You must also know how **erosion, transportation** and **deposition** change as the river moves from its **upper**, through its **middle**, to its **lower course**, and twice questions have been asked about this. You should base your explanation on the fact that the **energy** of the river decreases as its gradient lessens between the upper and lower courses.

On a couple of occasions questions about **rainfall in West Africa** have appeared in the Hydrosphere section. Answer these as you would if they were in the Atmosphere section where they would normally appear, referring to movement of the ITCZ and the mT and cT air masses.

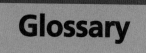

# Glossary

**Basin lag** – The time it takes for precipitation to make its way into the main river by run-off, tributary flow and throughflow of soil and groundwater.

**Discharge** – The volume of water in a river that passes in a given time – usually measured in cubic metres per second (cumecs).

**Helicoidal flow** – The corkscrew flow of the river as it moves downstream.

**Inputs** – Water which enters the drainage basin, i.e. all forms of precipitation.

**Outputs** – The water that leaves the drainage basin by transpiration, evaporation and rivers.

**Riffles** – Shallow areas of the river bed.

**Storage** – Water that is held in the basin, e.g. lakes, soil moisture, groundwater.

**Transfers** – The sideways movement of water through the basin, e.g. tributaries, throughflow, groundflow.

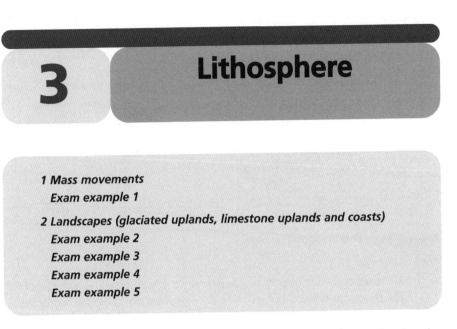

# 3     Lithosphere

*1 Mass movements*

   *Exam example 1*

*2 Landscapes (glaciated uplands, limestone uplands and coasts)*

   *Exam example 2*

   *Exam example 3*

   *Exam example 4*

   *Exam example 5*

Questions on the lithosphere relate to **mass movements** and the erosional and depositional processes that have produced the landscapes of **glaciated uplands**, **limestone uplands** and **coasts**.

If you are looking through past papers and find questions about scarp and vale landscape, don't worry – it has recently been removed from the syllabus and replaced by coasts. Also, the mass movements mudflow and soil creep have been removed and happily not replaced by anything.

## 1   MASS MOVEMENTS

At one time this topic occurred very frequently, and since 1996 has still cropped up every three or four years.

You need to have a detailed knowledge of **rockfall/scree formation** and **landslip/slumping**.

## EXAM EXAMPLE 1

This is a typical question.

> **The reference diagram (on the following page) shows two different types of mass movement on slopes.**

> **For each, *describe* and *explain* the conditions and processes which encourage it to take place.** 12 marks
>
> *Reference diagram. Selected mass movements on slopes*
>
>
> ROCKFALLS
> Bare rock
> Scree slope
>
> LANDSLIP (SLUMPING)
> Sandstone
> Clay

**Answer 11**   This is a good answer.

Landslip

The sandstone rock layer is porous (✔). When it rains heavily water soaks into the sandstone (✔) and cannot drain through it because the clay beneath is impermeable (✔). Consequently the sandstone becomes saturated and much heavier (✔). The clay is a soft rock (✔) and may not be able to support the weight of the saturated sandstone (✔) and slumps along a curved sheer plane (✔). When this happens the movement tends to be rapid (✔).

Rockfall

Frost shattering breaks rocks off cliffs (✔). They fall under gravity to form scree slopes (✔)                    Total: 10 out of 12

**Why is this a good answer?**

Up to 8 marks may be available for one mass movement and the landslip explanation is sufficiently detailed to receive those 8 marks. The rockfall explanation is weak, however, only receiving 2 marks. To get the additional 2 marks required the candidate could have referred to any of the following – processes of frost shattering and exfoliaton; location; climatic influences; shape of fragments; or influence of vegetation.

# 2 LANDSCAPES (Glaciated uplands, limestone uplands and coasts)

There are two main types of question, those based on OS maps and those that ask for explanations of how physical features are formed.

# Those based on OS maps

These tend to occur roughly every three years and, since there are three different landscapes in the syllabus, any one landscape type does not occur very frequently. However, the questions are straightforward and, in the case of glaciation, it is reinforcing what you have already learned at Standard Grade. So do not skip this area of revision.

You are likely to be asked to give map evidence that an area has either been glaciated or is an upland limestone area. Consequently, you must refer to specific features on the map and give grid references. **Marks are certain to be deducted if you fail to give grid references.**

Below are contour patterns and map symbols for a range of landscape features which you should know.

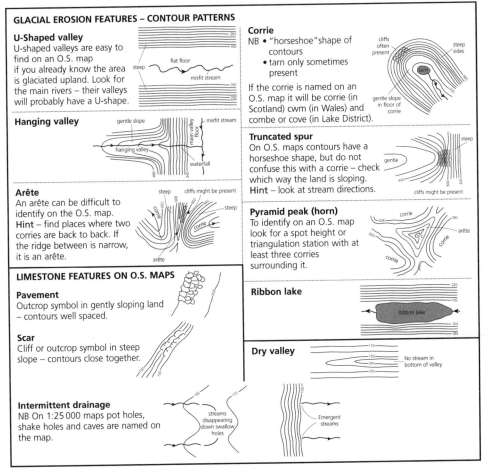

At the time of writing there has not yet been an OS map coast question. However, the following question is a strong possibility and something similar must appear soon:

> **Describe the physical features of the coastline between grid reference x and grid reference y.**

To answer this you should note the following:

- General shape – e.g. smooth, rugged, straight, indented, curved
- Trend of the coast – e.g. running north to south
- Specific features such as spit, headland, stack, wave-cut platform, cove, bay, beach. You must give grid references for these and, if possible, dimensions – e.g. a beach 500 m long and 50 m wide
- The form of the land immediately behind the coastline – e.g. cliffs, dunes, saltmarsh, lagoon, river mouth, steeply sloping up to 90 m.

Also, make sure you know the OS map symbols for marsh/salting, slopes, cliff, flat rock (wave-cut platform), dunes, sand, mud and shingle.

# EXAM EXAMPLE 2

> **Describe the physical features of the coastline between 260590 and 301620**
>
> 8 marks

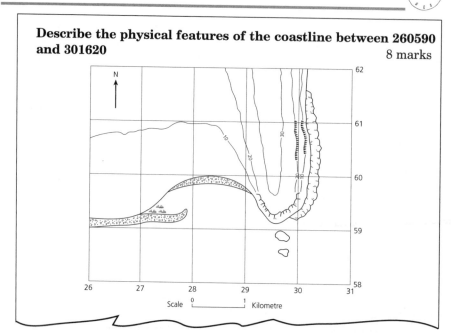

**Answer 12** **This is a very good answer.**

In the west the coast trends E – W, but at 300594 it swings round to run N – S (✔). In 2859 there is a bay (✔) with a shingle beach (✔). At 275592 there is a spit (✔) which indicates that longshore drift is towards the east (✔). There is a headland at 296592 (✔), on both sides of which are cliffs (✔) which are 20 m high (✔). In 2958 there are small islands that could be stacks or stumps (✔). On the east coast there is a wave-cut platform (✔), 2 km long and 200 m wide (✔). At 300605 there is evidence of slumping (✔). In the west the land is fairly flat but in the east it rises steeply to over 30 m (✔).                              Total: 8 out of 8

This is a good detailed answer which comfortably achieves full marks.

## Explanations of how physical features are formed

This is the most common lithosphere question and it is just as likely to be about coasts as it is to be about limestone or glaciation. It is well worth revising thoroughly because this knowledge can also be applied in the Rural Land Resources section of Paper 2.

> **Important tips:**
> - Always use **annotated diagrams** in your answer, even when not specifically asked for them.
> - **Practise** drawing the diagrams when you are revising.
> - Use simple line drawings – they are much quicker and easier    to draw than artistic block diagrams.
> - Give plenty of detail on the processes of erosion and deposition – do not be fooled into thinking: 'This is easy, I got full marks for this at Standard Grade'. There may be twice as many marks for a question like this at Higher.
> - When given a choice of features, choose the one you can write most about, e.g. a stack would be a better choice than a cave, because you can explain cave formation on the way to explaining the stack.

You may have learned more, but make sure you can at least explain the formation of the following features:

- **Glacial erosion** – U-shaped valley, corrie, tarn/corrie lochan, pyramid peak, arête, hanging valley, truncated spur, roche moutonnée, ribbon lake
- **Glacial deposition** – terminal moraine, drumlin, ground moraine/boulder clay, erratic
- **Fluvio-glacial deposition** – outwash plain, esker, alluvial fan

- **Coastal erosion** – wave-cut platform, wave-cut notch, cave, blowhole, arch, stack, stump, cove, headland, bay
- **Coastal deposition** – beach, spit, bar/lagoon, tombolo
- **Limestone** – pothole, cave/cavern, stalactite, stalagmite, pillar, pavement, clint, grike, gorge, swallow hole, shakehole, dry valley.

# EXAM EXAMPLE 3

> **With the aid of annotated diagrams, explain how a corrie is formed.**
>
> 8 marks

**Answer 13**   **This is a weak student answer.**

*Snow gathers in a hollow on a hillside and turns to ice (✔). It moves downhill and deepens the hollow by erosion (✔).*

Glacier moves downhill

**Why is this a weak student answer?**

This is an inadequate answer with no detail of erosion processes. The diagram also demonstrates two common errors. It has been carelessly drawn showing an unfeasibly overhanging backwall and the glacier does not vacate the hollow like a train leaving a station. Total: 2 out of 8.

**Answer 14**   **This second answer shows how it should be done.**

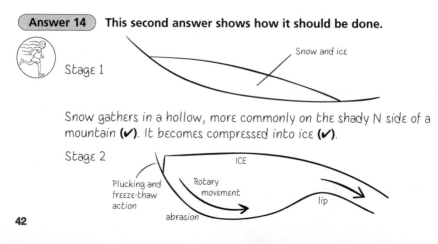

Stage 1

Snow and ice

*Snow gathers in a hollow, more commonly on the shady N side of a mountain (✔). It becomes compressed into ice (✔).*

Stage 2

ICE

Plucking and freeze-thaw action

Rotary movement

abrasion

lip

As ice builds up in a cooling climate (✔) its weight causes it to move downhill under gravity (✔). At the back of the hollow freeze-thaw action loosens rocks (✔), which become frozen into the glacier (✔) and plucked out as the ice moves away (✔). This steepens the back of the hollow (✔). The bottom of the hollow is deepened by abrasion (✔) as the rocks frozen into the base of the ice are scraped across the ground (✔). The rotary movement of the ice erodes the middle of the hollow most deeply (✔) so that a shallow lip forms (✔).

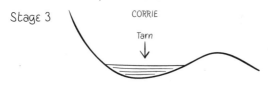

Stage 3     CORRIE

Tarn

When climate warms and ice melts a large armchair-shaped hollow or corrie remains (✔), which may contain a tarn (✔).

**Why is this a better answer?**

This is an excellent answer, more than worth full marks. It helpfully integrates diagrams and text, to clearly explain the processes in corrie formation. Note the simple but clear line diagrams. Total: 8 out of 8.

# EXAM EXAMPLE 4

> **Explain the formation of a sand spit, referring in detail to the processes involved.**      8 marks

**Answer 15**

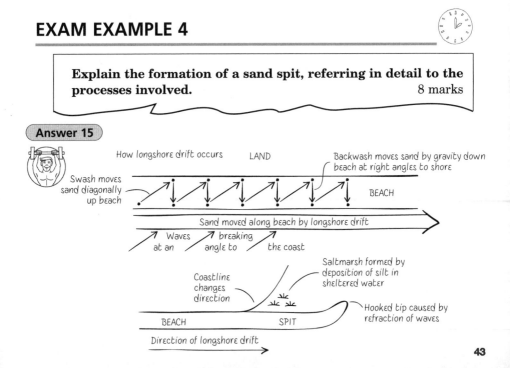

How longshore drift occurs     LAND     Backwash moves sand by gravity down beach at right angles to shore

Swash moves sand diagonally up beach

BEACH

Sand moved along beach by longshore drift

Waves at an breaking angle to the coast

Coastline changes direction

Saltmarsh formed by deposition of silt in sheltered water

Hooked tip caused by refraction of waves

BEACH     SPIT

Direction of longshore drift

This answer demonstrates the way in which full marks can be obtained entirely by providing well-annotated diagrams. Total: 8 out of 8.

# EXAM EXAMPLE 5

> **Explain the physical processes involved in the formation of stalactites and stalagmites.**      **8 marks**

**Answer 16**    **This is a good answer.**

These form in limestone caves (✔), where water is dripping from the ceiling (✔). The water contains dissolved calcium carbonate (lime) (✔). Every time a drip falls from the roof a small amount of water evaporates (✔) so that a tiny amount of lime is deposited on the ceiling (✔). This process is repeated with every drip (✔) so that gradually an 'icicle'-like stalactite (✔) grows downwards from the ceiling (✔).

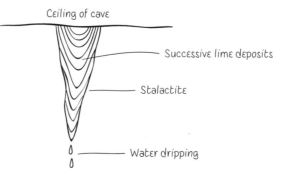

**Why is this a good answer?**

This is a good explanation of stalactite formation, but only gains 6 marks out of 8 because the candidate has omitted to explain stalagmites. He should have read the question more carefully.

# 4 Biosphere

*1 Soils*

*Exam example 1*

*2 Plant succession in sand dunes*

*Exam example 2*

*Exam example 3*

These questions will be either on **soils** or **plant succession in sand dunes** and are particularly predictable.

## 1 SOILS

Most commonly you are required to draw an annotated profile of just one of the three soil types and explain the processes (soil forming factors) that have created it. However, do not assume that you need only revise just the one soil type, because you may be asked to explain the differences between two soils or even refer to all three soils in an answer.

To help you there may be a diagram showing the main soil-forming factors or perhaps the soil profiles themselves. These can be useful prompts, reminding you what you should include in your answer.

Do not be confused by the profiles that may be drawn for you. Soils vary a lot so the profiles that appear in the exam may be slightly different to the ones you have been taught in school. Below are the profiles that so far have always appeared in the exam.

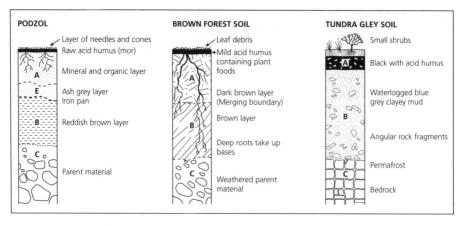

Do not be put off by the E horizon in the podzol. This is just the ash grey lower layer of the A horizon.

# EXAM EXAMPLE 1

**Study the Reference Diagram.**

**Select *one* of the following soil types:**
i    gley;
ii   podzol;
iii  brown earth.

**With the aid of an annotated sketch of a soil profile, *explain how* the major soil forming factors shown in the diagram have contributed to its formation.**                    12 marks

*Reference Diagram: Main factors affecting soil formation.*

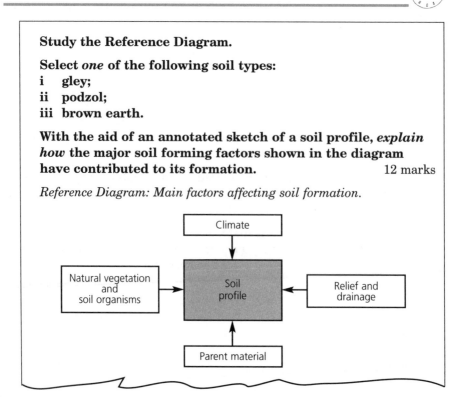

**Answer 17**  **This is a weak answer.**

*Podzol*

*A number of factors influence the way in which a soil is formed. They can be relief and drainage, climate, parent material, natural vegetation and soil organisms.*

*Bacteria helps the decomposition of plant matter. The soil water can move up or down in the soil so there might be leaching.*

*A podzol has a grey A horizon and a red-brown B horizon (✔), sometimes with a hardpan in between (✔). The soil forms beneath coniferous trees (✔) in a cold damp climate (✔), usually where the soil is well drained (✔). Strong leaching occurs in the soil (✔).*                    Total: 6 out of 12

**Why is this a weak answer?**

Paragraph 1 merely restates the question – 0 marks. Paragraph 2 contains a couple of general facts about soil formation but has not applied them to the formation of a podzol – 0 marks. Paragraph 3 contains six descriptive points about the soil, so the total is only 6 marks out of 12 which is a pity, because the candidate clearly knows the key facts about a podzol, but has neither linked them nor explained them. For instance, he could have linked the cool wet climate to downward movement of soil water to leaching, thus helping to explain why leaching occurs. Also, there is no diagram, so easy marks have been lost.

**Answer 18**  **This is a good answer.**

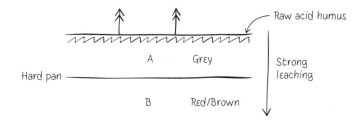

*Podzols occur in cool wet climates (✔) where precipitation exceeds evaporation (✔). Consequently water soaks down through the soil (✔) and leaches minerals from the A horizon and deposits them in the B horizon (✔). These minerals include iron oxide which gives the B horizon its reddish brown colour (✔). All that remains in the A horizon is silica which has a grey colour (✔). The soil usually occurs below coniferous forests (✔)*

which only provide hard needle shaped leaves providing few bases (✔) so the soil is acidic (✔). In the cold conditions there is very little bacterial activity to break down the leaves (✔) so there is a layer of raw acid humus at the surface (✔) called mor (✔). There are also very few soil organisms like worms in the cold conditions (✔). Therefore the soil does not get mixed and clear horizons develop (✔).          Total: 12 out of 12

**Why is this a better answer?**

The diagram, though lacking explanation, is worth a couple of marks and the written part of the answer logically explains how each part of the profile has been produced. Although there is no explanation of the hardpan and two of the boxes in the reference diagram have not been referred to, there is sufficient detail relating to climate, vegetation and soil organisms to amply compensate. Note also the use of explanatory link words such as 'consequently', 'so' and 'therefore'. This is a good answer which comfortably achieves full marks.

# 2  PLANT SUCCESSION IN SAND DUNES

> **Describe and explain the plant succession for a sand dune habitat. You should make reference to specific plants.**

> **Explain the changes in vegetation that take place across the sand dune system.**

> **Suggest reasons for changes in vegetation along the line of the dune transect. You should refer to a range of environmental factors.**

The above are just a few of the many ways this question has been posed and they all ask the same thing. Thus, you should be able to name at least one or two plants from each significant part of the dune system – embryo, fore-dune, yellow, grey and the slacks – and explain the way in which the plant is adapted to the conditions. If asked about the whole succession, as the questions above do, make sure you refer to all these parts.

Here is another example of this question type which focuses on three specific parts of the system.

# EXAM EXAMPLE 2

**Study the Reference Diagram.**

**Describe the plants likely to be found at sites A, B and C and give reasons why they occur at these locations.** 12 marks

*Reference Diagram: Transect across sand dune coastline.*

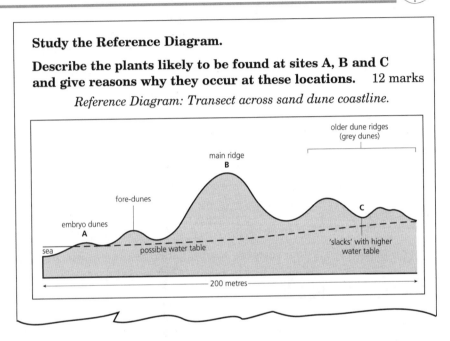

**Answer 19** **This is a weak answer.**

At A there are plants which can cope with difficult conditions.

At B marram grass grows (✔), which has long roots that can reach down to the water table (✔). It can grow upwards very fast, at one metre per year (✔) enabling it to keep pace with the build-up of sand (✔). It also has a curved shiny blade (✔) which cuts down water loss by transpiration (✔). Its flexibility enables it to bend in the wind and not break (✔).

At C the grey dunes have more humus, meaning that a wider variety of plants can grow, such as sand fescue and sand sedge.

**Why is this a weak answer?**

Paragraph 1 gets no marks, because the statement is too vague and no examples of plants are given. Paragraph 2 contains seven points worthy of marks, but there would be no more than 6 marks available for any one of the three locations. Paragraph 3 regrettably gains no marks because location C is in a dune slack and not the grey dunes. Hopefully the examiner would not be so cruel, but you should assume the worst! Total: 6 out of 12.

**Answer 20**    **This is a better answer.**

At A the conditions are very salty (✔) and the wind tends to dry plants out (✔). A plant that can cope with these conditions is Lyme Grass (✔) which also binds the sand together (✔).

At B Marram Grass grows (✔). This grows very fast to keep up with the accumulation of sand (✔). It has long roots (✔) to reach deep water below the ground (✔).

At C you are between dunes and close to the water table so the soil is wetter (✔) and reeds can grow (✔).

**Why is this a better answer?**

At first glance this may seem a fairly thin answer. However, it is concise and to the point and does refer to all three locations A, B and C. Total: 10 marks out of 12.

Roughly every three or four years the short question below is asked.

# EXAM EXAMPLE 3

> **Explain fully what is meant by the term 'climax vegetation'.**
>
> 6 marks

**Answer 21**    **So, here is a model answer – swot it up!**

Plant communities change as environmental conditions change (✔). At first there will be hardy pioneer species (✔) which will change the environment by providing shelter from wind (✔) and humus for the soil (✔), thus allowing different plant communities to develop in succession (✔). The final stage of these changes is known as climax vegetation (✔) when the plant community is stable (✔) and in balance with the environmental conditions (✔) of climate and soil (✔). A good example is the oak/ash forest which developed over lowland Britain (✔).

# Glossary

**Brown Forest soils** – another name for Brown Earth soils.

**Coastal dune transect** – a line drawn from the coast inland across a dune system.

**Parent material** – the rocks from which a soil has weathered.

**Plant succession** – the series of plant communities that develop as the environmental conditions change.

**Sand dune belt/dune system** – two different terms for the series of dunes which form in certain coastal areas.

**Soil biota** – Living organisms within the soil (e.g. worms, bacteria, roots, insects).

# 5 Population

**1 Population structure**

   *Exam example 1*

**2 Population change**

   *Exam example 2*

**3 Migration**

   *Exam example 3*

These questions fall into three general sub-topics: **population structure**, **population change** and **migration**. Migration and population structure questions occur roughly every three or four years and population change questions a bit more frequently, though in a wider variety of forms.

There is considerable overlap and linkage between each sub-topic, so that knowledge of one can be applied to questions on the others. If you know the following information you should be able to apply it to a number of different questions:

- The meanings of key terms such as Birth Rate, Death Rate, Infant Mortality Rate, GDP/GNP.
- Reasons for changing Birth Rate and Death Rate in the Demographic Transition Model and in ELDCs and EMDCs.
- The effects, both good and bad, of changing Birth Rates and Death Rates in both ELDCs and EMDCs.
- The reasons for and consequences of rapid population growth in ELDCs and slow or negative growth in EMDCs (this is closely linked to the previous two points).
- Methods of gathering population data and why ELDCs have more difficulty conducting a population census than EMDCs.
- Increasing Death Rate due to AIDS in many ELDCs and the consequences.

# 1 POPULATION STRUCTURE

These questions are based on age–sex **population pyramids**. These are basically bar graphs showing the population structure of a country either in number or as percentages at different ages for both males and females. You should be able to:

- using a pyramid, describe the population structure of a country, or contrast it with another pyramid, usually another differently developed country or perhaps the same country at a different date;
- identify the typical pyramids for an ELDC and an EMDC and understand the reasons for the different shapes and the consequences for these countries.

> **HINT:** When describing pyramids, do not be tempted into giving a lot of detail on the figures for each bar. Concentrate on the broad age ranges, like children (under 16), working age population (16–65) and the elderly (over 65).

# EXAM EXAMPLE 1

> **Study Reference Diagram.**
>
> a) *Describe* and *explain* the population structure of Botswana in *2000*.      10 marks
>
> b) *Describe* the population structure of *2025* and *suggest reasons* why it is expected to be so different. Note that one group is tracked on both diagrams.      8 marks
>
> *Reference Diagram (Age pyramids for Botswana in Southern Africa, 2000 and 2025)*
>
>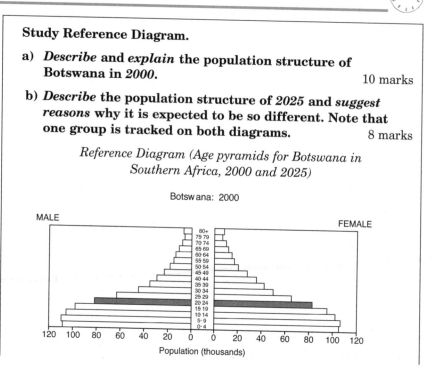

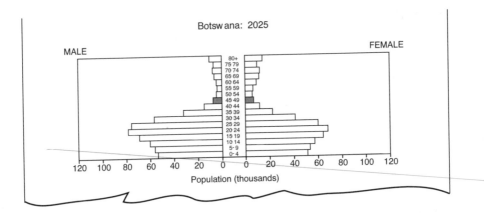

Botswana: 2025

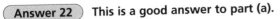

**Answer 22** **This is a good answer to part (a).**

(a)

The base of the pyramid is very broad with the largest number being those of children under 15 (✔). The largest number is in the 5–9 age group (✔).

The reason for this is that in Botswana there is a high birth rate due to a lack of family planning (✔). Parents often want large families because many children die young in their country and they want some to survive (✔).

The top of the pyramid tapers very quickly so that there are comparatively few people living into old age (✔).

This is because Botswana is a poor country, an ELDC (✔) and people die early because the country cannot afford to provide many doctors, hospitals and medicines to treat them when they fall ill (✔).

Total: 7 out of 10

**Why is this a good answer?**

There would be 4 marks available for description and in paragraphs 1 and 3 the candidate picks up three of these but fails to note the lack of people of working age, making the population heavily dependent. Paragraphs 2 and 4 provide 4 explanatory marks, but she could have offered further explanation of high birth rate by mentioning the need for children to help with work on farms and lend help to their parents in old age. High death rates are also due to poor diet and lack of access to clean water/poor sanitation leading to disease.

Note that specific knowledge of Botswana is not necessary to answer this question well. You are told that it is located in Africa, from which you can deduce that it is an ELDC.

**Answer 23**    **This is a good answer to part (b).**

(b)

The population in 2025 will be much smaller than in 2000 (✔). The largest percentages of the population are between 15 and 35 (✔). Birth Rate will decline (✔) since there are fewer children in 2025 than in 2000 (✔). There will be very few people over 40 (✔). In fact there will be more people over 65 than between 40 and 65 (✔).

In middle age the population has been hit by AIDS (✔), picked up by unprotected sex (✔). Also children can be born with AIDS, explaining the drop in child numbers (✔).

Total: 7 out of 8

### Why is this a good answer?

Lacking extensive knowledge of the reasons for the differences, this candidate has sensibly made sure she gets all she can for description, easily obtaining the 4 marks available. Mind you, she could also have emphasised the dramatic drop in number of the highlighted age group. To get the last explanatory mark she could have noted that the elderly might have escaped AIDS, being less likely to have had many sexual partners once the disease became widespread.

Note that the opening sentence is correct, because the pyramids are in actual numbers. Had they been in percentages it would not have been possible to make this statement.

If you do not have specific knowledge of the country concerned and you have to explain a big gap in its pyramid, as in this case, you will be credited with intelligent suggestions such as migration, natural disasters and wars.

# 2 POPULATION CHANGE

These questions often use as a source a **line graph** of a country's population or the **Graph of Demographic Transition** (or parts of it).

Although questions are posed in a variety of ways the wording is not difficult to understand and they usually boil down to understanding the reasons for the different birth and death rates in ELDCs and EMDCs and the advantages and disadvantages associated with the ways their populations are changing.

# EXAM EXAMPLE 2

a) **Study Reference Diagram A.**

   **Choose** *one* **of the stages from the demographic transition model. Describe and explain the factors which affect population growth at that stage.**                    8 marks

b) **Study Reference Diagram B.**

   **Describe the problems which can arise from falling birth rates in Developed Countries, such as Sweden.**     10 marks

*Reference Diagram A (Stages in the demographic transition model)*

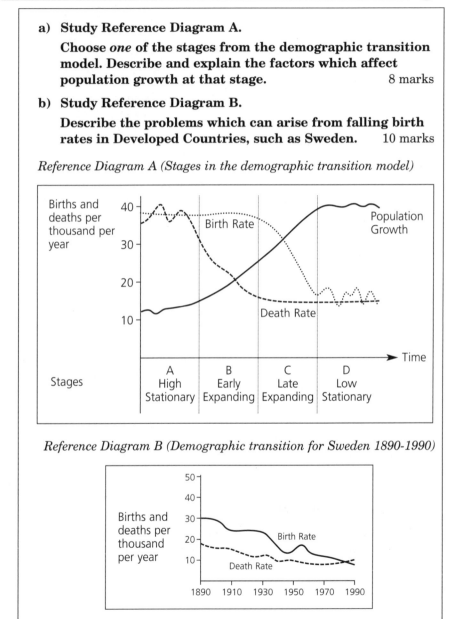

*Reference Diagram B (Demographic transition for Sweden 1890-1990)*

**Answer 24**    **This is an adequate answer to part (a).**

(a)

Stage B

The Birth Rate is high (✔) and the Death Rate starts high but is falling (✔). The Birth Rate is high because where there is a high rate of child death (✔) it is traditional to have large families (✔). Also there is a lack of contraception available (✔). The Death Rate is falling because of medical advances (✔).      Total: 6 out of 8

**Why is this only an adequate answer?**

This candidate has managed to get full marks for explanation, but he has only made two descriptive points at the start of his answer so missed some easy marks by not referring to the increasing population and not quoting any figures from the graph, e.g. 'Death Rate fell from 30 per 1000 to 16 per 1000.'

**Answer 25**    **This is a very good answer to part (b).**

(b)

As the Birth Rate and Death Rate get closer the rate of population increase will get less (✔). In Sweden the Death Rate has become slightly higher than the Birth Rate so the population will actually decrease (✔). This means that there will be less need for public services (✔) and hospitals and schools may have to close (✔) This could mean that some people would have to travel further to use these services (✔). Also, the number of elderly folk will become a larger proportion of the population (✔) and money will need to be provided by fewer workers (✔) to pay for their pensions (✔). Therefore the government may have to raise the retirement age (✔) or raise taxes (✔). The population will become more dependent (✔) and they might have to allow more immigrants into the country to do the work not being done by Swedes (✔).      Total: 10 out of 10

**Why is this a very good answer?**

This is not an easy question in which to achieve the full 10 marks so the candidate has done very well, clearly explaining how the population will become more 'dependent'.

# 3 MIGRATION

So far, exam questions have all concerned **international migration**, either **forced** or **voluntary**, and usually from an ELDC to an EMDC. However, the syllabus does include **rural–urban** migration. There seems no reason why this should not be questioned in future so make sure you revise this as well.

For each type of migration you should be familiar with one case study, knowing what the push/pull factors are, the benefits and problems created in the areas of both origin and destination, and any obstacles there might be to migration.

## EXAM EXAMPLE 3

> a) **With reference to an example of population migration between two named countries which you have studied,**
>   i) **explain how the migration was the result of a combination of 'push' and 'pull' factors, and**
>   ii) **suggest what sort of barriers or obstacles may have made the migration difficult.** 10 marks
>
> b) **Discuss the advantages *and* disadvantages which the migration has brought to**
>    ***either* the country of origin**
>    ***or* the country of destination.** 6 marks

**Answer 26** This is a good answer to part (a).

(a)

(i) Many people migrate from Mexico to the USA. Many of them slip across the border at night, but many get rounded up by border patrols and are sent back. However most of them try again and many eventually succeed. There are even road signs warning motorists to watch out for crossing immigrants.

The push factors from Mexico are low pay, poverty, poor education and poor health care (✔). The pull factors of USA are high pay, good standard of living, good education and good health care (✔).

Many migrants come from the Mexican countryside where there is not enough good land available to support all the population (✔). In USA there is a big demand for cheap labour (✔) to work at the harvests on the big farms and orchards in S California (✔). Because USA is a rich EMDC it can afford to provide good hospitals and many doctors and nurses which is a pull factor (✔).

(ii) It is difficult for the Mexicans to get into the USA because of immigration controls (✔) and many of those that do arrive are illegal immigrants (✔). They speak Spanish (✔) so may have difficulty communicating since the main language in USA is English (✔).

Total: 10 out of 12

### Why is this a good answer?

Paragraph 1 importantly names the example – if this had not been done it would have been marked out of 8. The rest of this paragraph, however, while true and, indeed, fascinating, is not answering the question. Make sure you stick to the point. In paragraph 2 there are two lists of factors and the second list is just the opposite of the first. At Higher you must develop your answer more fully, which is what the candidate goes on to do in paragraph 3. Additional obstacles that could have been mentioned are worries about racial discrimination and not wanting to be separated from one's family.

**Answer 27**   **This is a weak answer to part (b).**

(b)

The advantage to the USA is that low paid menial jobs will get done (✔), the sort that many Americans are not prepared to do, like fruit picking (✔). A disadvantage to USA is that housing has to be provided for all the immigrants (✔).

For Mexico, US dollars will be sent home to families (✔), but families find it hard to cope without a man about the house (✔). Total: 3 out of 6

### Why is this a weak answer?

The candidate has not read this question properly and has answered in relation to both country of origin and country of destination, so has not been credited with the marks in paragraph 2.

# Glossary

**Age pyramid/age–sex pyramid/population pyramid** – Three different names for a bar graph which illustrates population structure.

**Census** – A count of a country's population which may include a range of data such as age, sex, occupation, etc.

**Demographic Transition Model** – A combined line graph showing changing birth rate, death rate and population over time.

**Demography** – The study of population statistics.

**ELDC** – An economically less developed country, formerly known as a developing country.

**EMDC** – An economically more developed country, formerly known as a developed country.

**International migration** – The movement of people from one country to another.

**Losing country/donor country/country of origin** – Three terms used to describe the country from which migrants leave.

**Projected ageing** – The way in which a population may be predicted to become older.

**Pull factors** – Conditions in the country of destination which encourage people to migrate there.

**Push factors** – Conditions in the country of origin which make migrants want to leave.

**Receiving country/host country/country of destination** – Three terms used to describe the country to which migrants move.

**Rural–urban migration** – The movement of people from the countryside to towns and cities.

# 6 Rural Geography

1 Farming landscape/system

Exam example 1

2 Changes in the farming systems

Exam example 2

3 Population density

Exam example 3

Glossary

In your school or college course you will have examined case studies of each of three farming/agricultural systems: **shifting cultivation, intensive peasant farming** and **extensive commercial agriculture** (sometimes referred to as commercial arable farming).

For **each** of these you must be able to:

- describe and explain the main characteristics of the farming **system** and its landscape;
- describe and explain **changes** that have taken place;
- describe the **impact** of these changes on people and the environment;
- explain the relationship between a farming system and the **population density** of its area.

## 1 FARMING LANDSCAPE/SYSTEM

All three farming systems have been equally represented in this question, which occurs very frequently – twice every three years, although it may be worded in a variety of ways.

The question is most usually worded as '**Describe and explain the main features (or characteristics) of a farming landscape (or system)**'. However, the following are other ways the question has been posed and will require you to answer in much the same way as you would for the above:

> **Describe and explain the farming methods (or traditional methods) of a farming system (or type).**

> **Describe and explain in detail what is meant by one of the three farming types.**

> **Describe how various factors have affected farming activities of a particular system.**

The prompt for these questions is often a picture of the landscape, so make sure you can recognise each of the three landscapes.

Note that a farming landscape is closely related to the system of farming. Occasionally you are asked just to **describe** the farming landscape. In the case of extensive commercial agriculture, for example, this would mean mentioning the large area of the farms, the flat land, lack of hedgerows/walls/fences, the same crops in each field, large fields, big machines doing the work, big grain stores, grid-iron pattern of roads, sparse population, farms at regular well spaced intervals along roads, bigger settlements at junctions, etc. However, it is more common to be also asked to **explain**, which will inevitably lead you to talk about the farming methods/system as well, where they influence the landscape.

You may be provided with a reference diagram such as either of the models on the opposite page.

## Models of selected agricultural systems

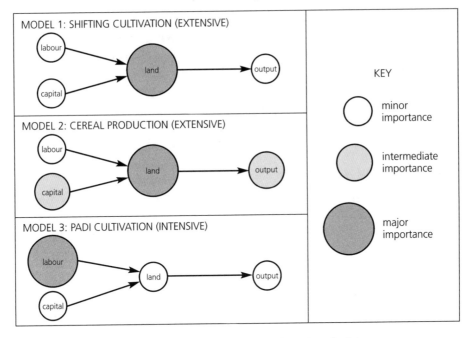

MODEL 1: SHIFTING CULTIVATION (EXTENSIVE)

labour
capital
land
output

MODEL 2: CEREAL PRODUCTION (EXTENSIVE)

labour
capital
land
output

MODEL 3: PADI CULTIVATION (INTENSIVE)

labour
capital
land
output

KEY

minor importance

intermediate importance

major importance

## Model of factors affecting farming decisions

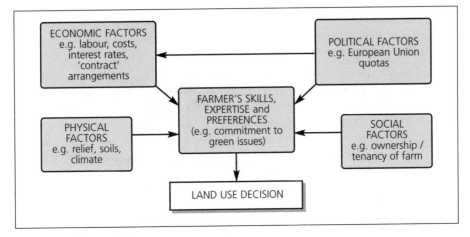

ECONOMIC FACTORS
e.g. labour, costs, interest rates, 'contract' arrangements

POLITICAL FACTORS
e.g. European Union quotas

PHYSICAL FACTORS
e.g. relief, soils, climate

FARMER'S SKILLS, EXPERTISE and PREFERENCES
(e.g. commitment to green issues)

SOCIAL FACTORS
e.g. ownership / tenancy of farm

LAND USE DECISION

These particular diagrams do not provide you with any answers, but they are a helpful prompt, reminding you what to include in your answer. Or, you may just be presented with a bare question such as the one on the following page.

# EXAM EXAMPLE 1

> **Choose *one* of the farming systems below and, referring to a named location, describe and explain the characteristics of the system.**
>
> i)    **Shifting cultivation**
> ii)   **Intensive peasant farming**
> iii) **Extensive commercial farming**        8 marks

**Answer 28**    **This is a very good answer.**

In the Canadian Prairies extensive commercial farming is practised. The soil is a fertile chernozem (✔) which, with the warm, dry summers and moderate rainfall (✔), is ideal for growing cereals (✔). The flat land enables machinery to be used for ploughing, sowing and harvesting (✔). Consequently field sizes and farms are very big (✔) and not many workers are needed (✔) so the population density is low, with widely spaced farmhouses in the landscape (✔). When the land is ploughed , the soil is easily blown away by the wind in the dry conditions (✔) so strip farming is practised (✔) leaving strips of fallow grass covered land to provide some shelter from the wind for the ploughed strips (✔). As a result only half the available land may be being cultivated, producing a low yield per hectare (✔).             Total: 8 out of 8

**Why is this a good answer?**

This is a very good answer which explains a number of characteristics, and yet the candidate has not mentioned monoculture, dry farming, the economies of scale, grain silo storage, availability of capital in a wealthy country, transport or other elements of the system. Clearly there may be some topics where you have more than enough to say for the marks that are available. So, watch the clock and do not get carried away.

# 2 CHANGES IN THE FARMING SYSTEMS

This sub-topic also occurs frequently – twice every three years on average. Intensive peasant farming has been the system most commonly tested, twice as frequently as the other two systems.

The question has been either (i) **Describe and/or explain the changes** which have taken place (or changes in crop output); and/or (ii) Discuss **the impact** of the changes on the people and/or way of life and/or farming landscape and/or environment. This has also been worded as 'Comment on the successes and failures of the changes'.

The most useful sources that may be provided for you are graphs showing changing crop yields or photos/sketches showing landscapes before and after change. However it may just be a quote, which does no more than lead you into the question or, as in the question below, a location map, which does no more than remind you where the particular farming system is practised.

# EXAM EXAMPLE 2

---

Study Reference Map.

Both shifting cultivation and intensive peasant farming have been subjected to change in recent years.

For *either* shifting cultivation *or* intensive peasant farming, and using named examples,

i)   outline the changes which have taken place, and

ii)  assess the impact these changes have had on the people, their ways of life and the landscape.                    14 marks

*Reference map. Generalised distribution of selected agricultural systems*

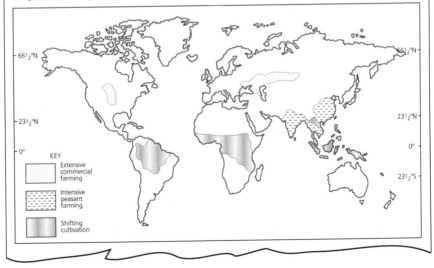

**Answer 29**    **This is a passable answer.**

(i)

Intensive Peasant Farming in India has traditionally been subsistence cultivation of rice. This is possible because rice needs high T's of over 20 C, high rainfall to flood the padis and a dry period for growing and they get this from the monsoon climate. The soil is fertile on the Ganges Plain and replenished with alluvium every time the river floods. Therefore there is a high yield per hectare and it means that many people can live on this land giving a high population density. The land is farmed intensively.

Big changes have taken place, called the Green Revolution (✔). HYV seeds and chemical fertilisers are now used (✔) which has doubled the crop yields (✔). New irrigation schemes have been developed in the Punjab (✔) so that two or three crops are now harvested each year (✔).

(ii)

With the increase in crop yields some farmers have not only enough to feed themselves, but a surplus that they can sell as cash crops to make money (✔) and India as a whole no longer has to import grain (✔). The general standard of living has risen (✔).        8 out of 14

**Why is this only a passable answer?**

Here is another candidate who knows a lot of facts about the topic, but has started by spewing out all he can remember. He has not taken time to read the question carefully and consider which facts are pertinent to this particular question, in this case farming changes and their impact. Paragraph 1 describes the traditional farming system, which is irrelevant and consequently receives no marks.

He gets on track in paragraphs 2 and 3, but has used up valuable time writing paragraph 1 and perhaps not had time to do himself justice once he has realised how he should be answering.

Important points that he has omitted are the increased use of **machinery**, which eases workload, but has a high cost of maintenance; **land reforms** which have consolidated fields and amalgamated farms into larger units, making farming more efficient, but forcing many people to migrate to the cities; the **high costs** of water pumps, fertilisers, pesticides, HYV seed and machinery, forcing small farmers into **debt**, which they cannot repay if crop yields or prices are less than expected and forcing them to sell up to the bigger landowners, who become increasingly wealthy; better **storage** reducing loss of grain to pests and damp; **new roads** to distribute surplus, improving communications in the area.

# 3 POPULATION DENSITY

This question has only been asked three times and never from 1999 to 2006. So, watch out, it could be appearing again soon!

# EXAM EXAMPLE 3

**Study Reference Table.**

**Choose *one* of the farming systems listed in the Table and, referring to an area you have studied, explain the population density associated with that farming system.**   8 marks

Reference Table (Farming systems and population density)

| *Farming System* | *Population Density* |
| --- | --- |
| Shifting cultivation | Low |
| Intensive peasant farming | High |
| Extensive commercial farming | Low |

**Answer 30**   **This is an excellent answer.**

Shifting Cultivation

The Boro Indians of the Amazon rainforest are shifting cultivators (✔), who cut down small clearings in the forest to enable them to cultivate the soil (✔). However, this exposes the soil to the heavy rain which soon leaches out the nutrients (✔). Consequently, after a few years the soil becomes infertile (✔) so that the Boro have to move on to another area of the forest and start again (✔). It takes a long time for the forest to recover and the soil to regain its fertility (✔), so not many people can successfully live in a rainforest (✔), especially since they also need undamaged forest around their clearings for hunting and gathering of fruit to add to their food supply (✔)     Total: 6 out of 6

**Why is this such a good answer?**

This answer clearly explains the underlying reasons for the necessity for a low population density.

The question did not ask for specific examples, but the candidate has quoted the example of the Boro and has been credited with a mark for this. It is always good practice to give examples. It lends authority to an answer, creating an image of an able candidate, so that the marker may be more likely to credit the candidate when making marginal points elsewhere in the exam paper.

# Glossary

**Farming landscape** – The appearance of the land under a particular farming system, including the shape of the land, settlement and road patterns, arrangement of fields, buildings, farm size and crops.

**Hectare** – A unit of area, 10 000 square metres.

**Human/economic inputs** – Elements which involve the expenditure of labour and money on a farm; e.g. fertiliser, pesticides, machinery, seed, etc.

**Outputs** – What is produced on a farm.

**Physical inputs** – The natural conditions which exist on a farm and which help to dictate what the farmer can do with his land; e.g. soil, climate, drainage, slope, etc.

**Yield** – The amount of a crop which is produced.

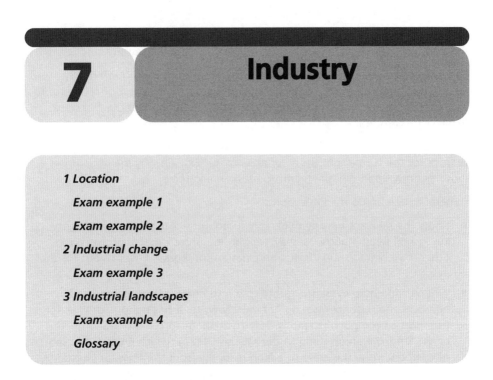

# 7 Industry

*1 Location*

   *Exam example 1*

   *Exam example 2*

*2 Industrial change*

   *Exam example 3*

*3 Industrial landscapes*

   *Exam example 4*

   *Glossary*

The majority of these questions relate to the **location** of industry, but there are also a variety of questions on industrial **change** and a few on industrial **landscapes**. All the key ideas were introduced at Standard Grade so you should be very knowledgeable on this.

## 1 LOCATION

This occurs very frequently. More than 80% of the exam papers have a question on this. It may be based on an OS map or a case study. The questions based on an OS map or other map sources are usually straightforward if you know the location factors for old (i.e. nineteenth century) and modern industry. It is just a matter of applying that knowledge to the particular map example.

The most common wording is 'Using map evidence describe and explain the **physical** and **human** factors that encouraged industry to locate in an area'. Consequently, it is important that you understand the difference between physical and human factors. Physical factors are those relating to landscape (e.g. site, rivers, deep water, mudflats for reclamation) and possibly climate. Human factors include anything man-made (e.g. railways, main road links, docks, canals) or otherwise to do with people (e.g. labour supply, market, EU subsidies, grants and loans).

Sometimes you are asked to describe the advantages of a **site** for industry. Strictly speaking, site is the land on which the industry is built, but reference to factors such as transport links and proximity to market and labour force are likely to be credited.

When asking about old industry the wording might be 'Suggest geographical reasons for the former prosperity of an area' or 'Explain why manufacturing industry grew up in the nineteenth century in an area'.

General advice on OS map questions:

- Know the relevant **map symbols** so you do not waste time consulting the key – spoil heap, quarry, railway, freight line, river, canal, motorway and other road symbols. Wks is an abbreviation for works, so watch out for that as well.

- When giving map **evidence**, quote grid references, place names, road numbers and distances. So, do not merely write 'It has good road links for transport of components and is close to a labour force.' Instead write 'It is only 1 km from an on-ramp to the M56 at 543271, which enables transport of components and is close to a labour supply from the town of Chester.'

- Make sure you **explain points**. Even if the question only asks 'Describe the advantages of the site', you cannot properly do this without making explanatory points.

- Do not write **'lists'**, which there is a temptation to do in map questions, e.g. 'There is flat land, open space and motorways.' Instead write 'The factory is built on flat land which is easy to build on, with open space beside it into which it could expand. There is good access to motorways, the M62 and M60, which will enable the products to be distributed easily.'

- Know what to look for on the map to identify whether it is an **old** or **new** industrial area (see the table on the next page).

*Map evidence for old and new Industrial Landscapes*

| OLD | NEW |
|---|---|
| Nearer the centre of city than the edge | Near edge of city |
| Large, but often irregularly shaped building outlines | Large rectangular blocks of buildings |
| Railway freight lines and sidings, canals, rivers, docks | Beside or close to A-class roads or near motorway junctions |
| Spoil heaps | Open space, room for parking and possibly woodland nearby |
| Close to C19 housing as shown by grid-iron street pattern | Nearest housing is C20 as shown by curvilinear street pattern |

# EXAM EXAMPLE 1

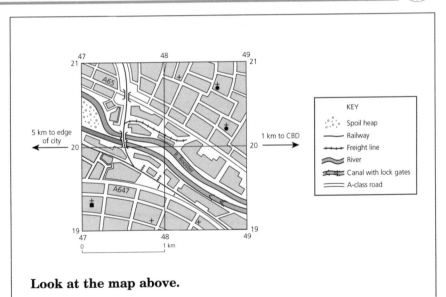

**Look at the map above.**

**Using map evidence, describe and explain the physical and human factors which led to industry locating in this area in the nineteenth century.**

8 marks

**Answer 31**    **This is a weak answer.**

There would have been raw materials nearby such as coal or iron ore. It would have provided employment in the area, causing living standards to rise. There is a canal, river and railway (✔).              Total: 1 out of 8

## Why is this a weak answer?

In the first sentence there is no map evidence of this, though it may well have been the case. In the second sentence the candidate has misunderstood the question by writing about the effects of the industry on the people nearby, rather than the factors leading to industry locating there. The third sentence is a list with no explanation.

**Answer 32**    **This is a much better answer.**

It is beside a river (✔), which would have provided a water supply used in the industrial processes (✔). There is a canal (✔), which would have been a cheap way to transport heavy raw materials such as coal and iron ore to the factories (✔). The railway and freight lines (✔) for example at 478203 (✔), would also have been used to transport raw materials and finished products (✔). The land is flat so easy to build on (✔). The old built up area would have provided a supply of workers (✔), who could have walked to their work since the housing is beside the industry (✔).

Total: 8 out of 8

## Why is this a much better answer?

This answer clearly explains each location factor. Note that in such a restricted map area there is less necessity to quote grid references.

Rather than interpreting a map, you may be asked to demonstrate knowledge of a case study.

# EXAM EXAMPLE 2

a) *"Traditional industries were often located on or near raw material sources"*

For any industrial concentration in the EU which you have studied, *describe* the *physical* factors which led to the growth of industry before 1950. **6 marks**

b) **Study Reference Map.**

*Describe* and *explain* the *human* factors which have led modern industries to locate along the corridor between London and Peterborough. **8 marks**

*Reference Map (A growing industrial area of the UK)*

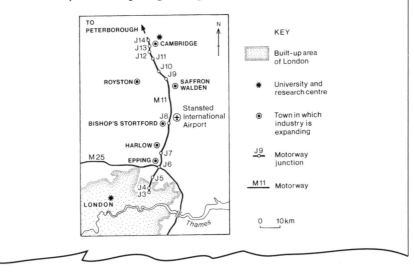

(a)

The Ruhr industrial area is located on the biggest coalfield in Europe (✔). Iron ore was mined in the Sieg Valley (✔) and limestone was also available in the area, so that all the raw materials necessary for smelting iron and steel were present (✔). At first coal was easily mined by open cast and adit methods on the exposed coalfield (✔). Once this was exhausted mining moved north on to the concealed coalfield (✔).

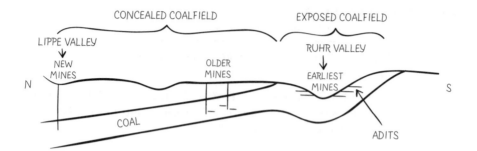

CONCEALED COALFIELD    EXPOSED COALFIELD

LIPPE VALLEY           RUHR VALLEY

NEW MINES      OLDER MINES      EARLIEST MINES

N                                           S

COAL

ADITS

After the local iron ore was worked out Swedish ore could be imported by barge up the navigable R Rhine (✔) to Duisburg, the major river port of the Ruhr (✔).          Total: 6 out of 6

## Why is this a very good answer?

This is an excellent answer that fully answers the question. The sketch map would also have received marks had the candidate needed them.

---

**Answer 34**    This is a weak answer to part (b).

(b)

London is nearby and there is a motorway, university and airport (✔). The R Thames enables goods to be imported and exported easily and there is flat land for building. The climate in the S of England is warmer than the N so firms would prefer to locate there.          Total: 1 out of 8

## Why is this a weak answer?

The first sentence is just a list of factors lifted from the reference map, with no explanation, so only receives one mark. The rest of the answer gains no more marks because the factors referred to are physical and the last sentence is fairly doubtful anyway.

---

**Answer 35**    Here is a much better answer.

(b)

Modern industries like electronics are light industries so the components and raw materials are likely to be transported by road (✔). The M11 motorway runs through the centre of this corridor (✔), providing a fast transport link for bringing in these materials and distributing the finished products (✔), because it also links with the M25 (✔). There are lots of junctions in this section of motorway (✔), so industries will have easy access to it (✔). Stanstead airport will be handy for management to travel abroad on business (✔). There are two universities nearby which will provide highly skilled people (✔) and they could be involved in research projects with the industries (✔). It is close to London, Britain's biggest city (✔), so this is a huge market (✔). There are a number of towns in the corridor, like Saffron Walden and Bishop Stortford (✔), which will provide a workforce (✔).                    Total: 8 out of 8

### Why is this a much better answer?

Notice how this candidate has used the reference map to identify important factors and then fully explained each one, intelligently developing her answer.

## 2 INDUSTRIAL CHANGE

On average there is a question on this every three or four years. You need to have detailed knowledge of a case study including:

- the reasons why the traditional industries in the case study area have declined;
- the social, economic and environmental consequences of that decline;
- how new industries are attracted to the area.

The source for this question may well be a table showing employment figures over a period of years and is likely to show a decrease in the primary and secondary industries and an increase in tertiary industry.

## EXAM EXAMPLE 3

**Study Reference Map.**

*Many industrial concentrations in Europe have experienced recent change.*

**With reference to *one* named industrial concentration in the European Union, describe and explain the recent industrial changes and discuss their impact on the area.**

Reference Map (Selected industrial concentrations in the EU)

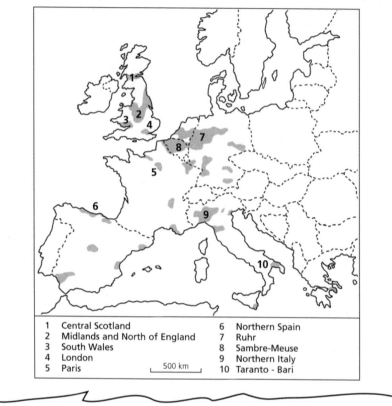

| | | | |
|---|---|---|---|
| 1 | Central Scotland | 6 | Northern Spain |
| 2 | Midlands and North of England | 7 | Ruhr |
| 3 | South Wales | 8 | Sambre-Meuse |
| 4 | London | 9 | Northern Italy |
| 5 | Paris | 10 | Taranto - Bari |

500 km

**Answer 36**   **This is a good answer.**

In the Ruhr the old traditional industries went into decline (✔). The increased use of machinery in the coal mines (✔) led to many miners losing their jobs (✔). The steel industry also declined due to competition from smelters in coastal locations (✔) which did not have the transport costs of the Ruhr (✔), which was suffering from industrial inertia (✔). Chemical and textile industries also declined, adding to unemployment (✔).

To solve unemployment the German Government has set up incentives for new industries to move to the region (✔) by providing grants and loans to the new companies (✔). Electronics and telecommunications are now major employers (✔). Veba makes mobile phones (✔) and the Japanese firm Hitachi makes machine tools in Krefeld (✔).          Total: 10 out of 12

**Why is this a good answer?**

This candidate would appear to have gained full marks since she has made 12 points. She has described and explained the changes very well and quoted some relevant examples. However, she has not answered the whole question, so cannot receive full marks. She has forgotten to discuss the impact of the changes, where she might have mentioned decreasing pollution, the utilisation of old brownfield sites, the different industrial landscape and the social problems related to the increase in unemployment.

# 3  INDUSTRIAL LANDSCAPES

This question is less frequently asked than it used to be, although it did occur in 2006. You need to know the characteristics of both an old industrial landscape and a modern industrial landscape. You will have notes on these at both Standard Grade and Higher so there is no excuse for not knowing this!

On three occasions the question has been based on the OS map and has simply asked you either to describe the industrial landscape of one area or the differing landscapes of two areas. The way to approach this is to first identify whether it is an old or new landscape by studying the map evidence (see table on page 75). Then describe what the landscape would be like. Obviously you can use what is shown on the map, but, although the map will not show them, you can also write about such things as the appearance of the buildings, housing types and environmental quality of the area.

# EXAM EXAMPLE 4

> **Look at the map on page 75.**
>
> **With the aid of map evidence, describe the likely industrial landscape of the area.**
>
> 6 marks

**Answer 37**   **This is a very good answer.**

There is a canal with lock gates (✔) at 486194 and 487193 (✔). Also there is a spoil heap at 471204 suggesting that this is an old industrial landscape (✔). The river and canal fronts are lined with big industrial works (✔), which would probably be several storeys high and have chimneys and few windows (✔). They might be steel works since there are freight lines leading off the main railway (✔). The housing is probably tightly packed tenements or terraces (✔), with very little open space or gardens (✔). The industries may well be polluting the air with smoke (✔) and the river with effluent (✔). There would also be considerable noise from the works, the railway and the busy main roads, A65 and A647 (✔).

Total: 6 out of 6

**Why is this a very good answer?**

This answer successfully combines map evidence with knowledge of the characteristics of an old industrial landscape. It easily achieves full marks.

# Glossary

**Brownfield site** – An area of derelict land that might be used to develop new industry.

**Greenfield site** – An area of open land, probably farmland, to be used for industrial development.

**Industrial inertia** – When an industry remains in a particular place despite the loss of the original advantages for its location there.

**Regeneration** – The improvements made to an area where traditional industry has declined.

# 8 Urban Geography

1 Urban land use zones/environments

Exam example 1

Exam example 2

Exam example 3

2 Urban change

Exam example 4

3 Site and location/situation

Exam example 5

Glossary

The two principal themes examined are **urban land use zones** and **urban change**, though other topics such as site and location do occasionally occur. At least every second year the question is based on an OS map. An accomplished Standard Grade pupil would be capable of tackling many of the questions, but remember again that much more detail is required at Higher Grade.

## 1 URBAN LAND USE ZONES/ENVIRONMENTS

The zones you should be familiar with are:

- The Central Business District (CBD)
- Nineteenth century housing
- Twentieth century housing
- The inner city (usually a mixture of C19 housing and industry and referred to in some texts as the 'twilight zone')
- Nineteenth century industry

- Twentieth century industry
- Old settlements swallowed by urban sprawl.

For each of these zones you should be able to:

- Identify the zone from map evidence
- Describe its characteristics and environmental quality, both from map evidence and your knowledge of these zones
- Explain its location within the city.

In the overwhelming majority of cases this question has been based on an **OS map**. 70% of such questions relate to **residential areas** (i.e. housing) and 50% to the **CBD**. Although you may have to identify them, detailed questions about industrial areas are more likely in the Industry section and the map evidence for these is summarised in the table on page 75. The table below summarises the map evidence for the other areas. Remember, however, if you are asked to describe the **environment** or **urban landscape**, you should also refer to the characteristics of the area that you will have learned – the types and storey heights of buildings, presence of gardens, air quality/pollution, noise, urban decay/regeneration, safety, etc.

### Map Evidence

| CBD | C19 HOUSING | C20 HOUSING |
|---|---|---|
| Centre of city | Near CBD | Edge of town |
| No open space | Very little open space | Open spaces |
| Tightly packed buildings with high street density | Grid-iron pattern of streets | Curvilinear street patterns and cul-de-sacs |
| A-class roads converge | Close to old industrial areas | Open spaces separate it from industrial areas |
| Many churches | Many churches | Fewer churches |
| Railway and bus stations | Main roads and possibly railways and canals | More minor roads |
| Historic buildings, e.g. cathedrals, castle | NB You can identify older settlements swallowed by urban sprawl by evidence of a small CBD fairly far from centre of city | |
| Tourist information centre and museums | | |
| Important public buildings, e.g. town hall | e.g. convergence of A-class roads, cluster of churches, town hall | |

# EXAM EXAMPLE 1

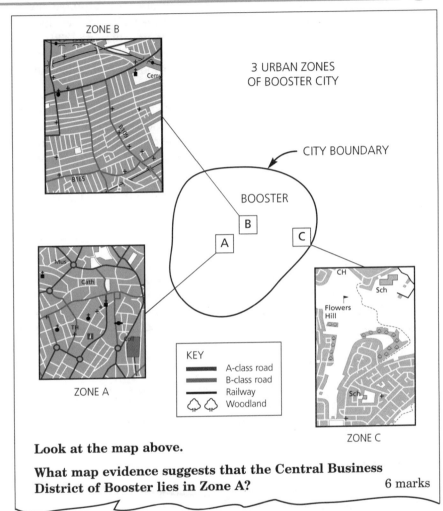

ZONE B

3 URBAN ZONES
OF BOOSTER CITY

CITY BOUNDARY

BOOSTER

B

A

C

KEY

▬▬▬ A-class road
▬▬▬ B-class road
──── Railway
♧ ♧ Woodland

ZONE A

ZONE C

**Look at the map above.**

**What map evidence suggests that the Central Business
District of Booster lies in Zone A?** 6 marks

---

**Answer 38** **This is a good answer.**

There is a high density of streets (✔) and very little open space (✔) due to
the demand for building land. The town hall is in this area (✔) and it is
very accessible, since there is both a bus station (✔) and a large railway
station (✔). There is a tourist information centre (✔) and a cathedral (✔)
as well as many churches (✔). Total: 6 out of 6

**Why is this a good answer?**

The only weakness in this answer is that the candidate has not quoted any grid references, for which two marks would have been available. However he has given more than enough examples of features that are typical of a CBD and has not been penalised for the lack of grid references due to the limited area of the map on which the question focuses. Do remember though that, when asked for map evidence, it is always advisable to give grid references.

This is in fact a particularly easy question – you will not encounter any easier ones! Since no explanation is required, only map evidence, this is a rare occasion that one might get away with writing a list, as the candidate has done in the last sentence. But, do not be tempted – keep the habit of writing developed points, as he does in the first two sentences.

# EXAM EXAMPLE 2

> **Look at the map above, on page 87.**
>
> **Zones B and C contain contrasting residential environments. Describe the residential environments of Zone B and Zone C and suggest reasons for the differences.**　　12 marks

**Answer 39**　　**This is a weak answer.**

Zone B is busier and dirtier than Zone C and the street patterns are different. The air is less polluted in Zone C (✔). The buildings are older in Zone B (✔). There is more open space in Zone C (✔). There are more churches in Zone B (✔).　　Total: 4 out of 12

**Why is this a weak answer?**

It is sometimes difficult to decide how little to accept to award a mark and maybe the marker has been generous on this occasion! However, this is clearly a weak answer, made up of simple comparative statements without any attempt to suggest reasons for the differences. The points made in the first sentence are too vague, but the remaining sentences do gain credit, even though each statement is the bare minimum necessary to obtain a mark. You should always attempt to include more detail and develop your answers as the following answer demonstrates.

**Answer 40**    This is a better answer.

Zone B

Here the streets have a grid-iron pattern (✔) and the houses are probably
C19 tenements or terraces (✔), packed close together to house workers in
the C19 factories (✔). There is very little open space (✔) and there would
be a lot of noise and fumes (✔) near to the railways and busy A class and
B class roads (✔).

Zone C

Here there is much more open space (✔) with woods and a golf course (✔)
so the air is much cleaner than in Zone B (✔). The streets have C20
geometric curved patterns with cul-de-sacs (✔) to restrict traffic to the
local residents (✔) and consequently make the streets safer for children (✔).

Total: 12 out of 12

**Why is this a better answer?**

This is a competent answer which just achieves full marks. Notice that the
candidate has combined both map evidence (street patterns, roads, railway, golf
course, open space, woodland) with knowledge of these environments (housing
types, air quality, noise, traffic) and has offered some reasons for the differences
as well. Failure to give reasons would have meant being marked out of 10.

The candidate could also have referred to the different number of churches in
the two areas and the types of houses and presence of gardens in Zone C.

On two occasions a land use question has required knowledge of a case study,
such as the question below.

# EXAM EXAMPLE 3

**Study Reference Diagram.**

**The diagram (on the next page) shows selected changes in
land use from the Central Business District to the edge of a
typical city in the *Developed* World.**

**With reference to the city you have studied, *describe* and
*explain* the differences in land use from the centre of the city
to the edge.**                                                    14 marks

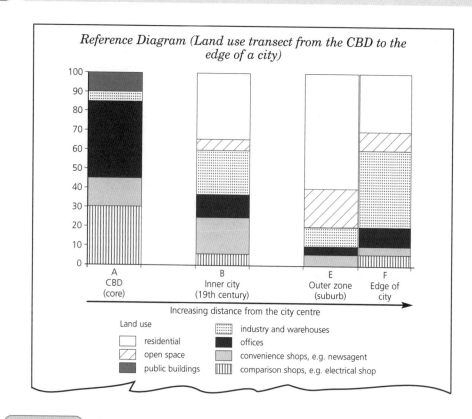

Reference Diagram (Land use transect from the CBD to the edge of a city)

Increasing distance from the city centre

Land use
- residential
- open space
- public buildings
- industry and warehouses
- offices
- convenience shops, e.g. newsagent
- comparison shops, e.g. electrical shop

**Answer 41** This is a very good answer.

Edinburgh – from CBD to west edge of city.

The CBD is centred on Princes Street (✔) around which there are many shops and offices (✔) due to this part of the city being the most accessible (✔), with A class roads meeting there (✔) and St Andrew's bus station (✔) and Waverley train station being located there (✔). There is no residential land use here due to the high rental costs, which only large companies and big department stores can afford (✔). Since this is also the oldest part of town there are historic buildings such as the castle (✔). Also Edinburgh is Scotland's capital so the Scottish Parliament is located here (✔).

The Gorgie and Dalry areas are in the Inner City (✔) and contain a mixture of old, often redeveloped, industrial buildings and tenement housing (✔), built in the nineteenth century to house the factory workers and from where they would be able to walk to their work (✔). There are railway lines and the Union Canal (✔), built to transport raw materials to the old factories (✔). There are many low order services such as newsagents to cater for the local residents (✔).

Further out is Corstorphine (✔), a predominantly residential area. It is very desirable with large detached and semi-detached houses (✔). They have large gardens and there is a zoo and golf course so there is plenty of open space (✔).

On the west edge of the city in Sighthill there is modern council housing and tower blocks (✔). There is also an industrial estate taking advantage of a cheap greenfield site (✔). The Gyle Shopping Centre and Edinburgh Business Park (✔) are located at the junction of the ring road and the M8 motorway (✔) where there is easy access for cars (✔).

Total: 14 out of 14

**Why is this a very good answer?**

Much more might have been said in this answer and yet the candidate easily achieves full marks. He has referred to each of the zones identified in the reference diagram and explained each of the land uses he has described – see if you can identify the descriptive and the explanatory points. 4 marks would have been available for naming appropriate places and again he has easily achieved these. Had he not named any then the answer would have been marked out of 10.

# 2 URBAN CHANGE

Every three years on average there is a question on this. So far, it has always focused on change in either the **CBD** or the **Inner City**. You need to have detailed knowledge of a case study.

You should know:

- The way the city has changed
- The reasons for the changes
- Specific named examples of changes
- How successful the changes have been.

# EXAM EXAMPLE 4

Study Reference Map which shows land use zones in Liverpool. Choose *one* of the land use zones A or B identified in the key.

Referring to Liverpool, or any other city you have studied, describe and explain the changes which have taken place in your chosen zone in recent years.          10 marks

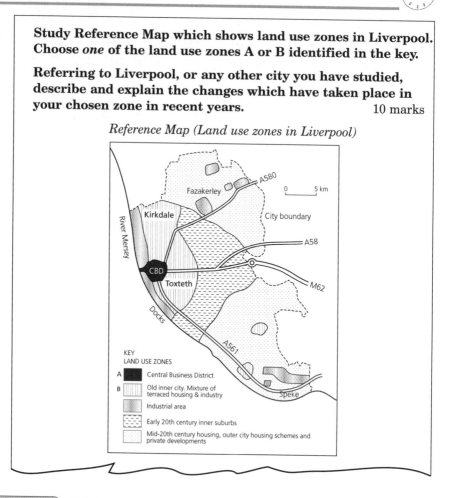

*Reference Map (Land use zones in Liverpool)*

KEY
LAND USE ZONES

A   Central Business District
B   Old inner city. Mixture of terraced housing & industry
    Industrial area
    Early 20th century inner suburbs
    Mid-20th century housing, outer city housing schemes and private developments

**Answer 42**   **This is a weak answer.**

B Old Inner City

In recent years the Inner City has changed considerably. Old industries have closed down (✔) and old housing has been demolished or renovated (✔). Many buildings have been cleaned up by sandblasting (✔). People have moved out to new towns (✔). Derelict land has been landscaped (✔). Old canals have been converted to inner city motorways (✔).

Total: 4 out of 10

## Why is this a weak answer?

This candidate has either not read the question properly, failing to realise that she can refer to a city that she has studied, or she does not know much about Liverpool. The last sentence is not true of Liverpool and she may be thinking of Glasgow. There is no explanation and no examples have been given of specific developments in particular named areas of Liverpool or whichever city she is talking about. Although she has made six points they are all descriptive, so she can only achieve a maximum of 4 marks. And, incidentally, the first sentence is unnecessary.

**Answer 43** **The following answer is much better.**

A Central Business District of Aberdeen

In the main street, Union Street, some large shops have closed down (✔) due to competition from large out of town shopping centres (✔) at Bridge of Dee, Bridge of Don and Portlethen (✔), which are more accessible by car than the congested town centre (✔) and where land rentals are cheaper on greenfield sites (✔). There are, however, new shopping centres in the CBD as well, such as the Trinity Centre and Bon Accord Centre (✔), which have the advantage of being under cover so that shoppers can go from one store to another without being affected by the weather (✔). Because of the increased amount of traffic there is considerable congestion (✔). Therefore a new road has been built under Union Street (✔) and right turns are now prohibited at many junctions on Union Street to keep the traffic moving more smoothly (✔). Large multi-storey car parks like the one in Harriet Street (✔) have had to be built, sometimes connected to the shopping centres (✔). As a result of the oil industry Aberdeen has a lot of wealth, so there is a big demand for entertainment and pubs (✔). Also not so many people attend the churches as they used to (✔) so some of them have been converted into clubs and pubs like The Ministry (✔). In order to make the street more pleasant and safe for pedestrians the lower part of George Street has been closed to traffic (✔).    Total: 10 out of 10

## Why is this a much better answer?

This candidate has read the question properly and wisely chosen to write about a city which she obviously knows well, enabling her to give plenty of specific examples. She has also explained the changes – note the use of linking words like 'due to', 'so that', 'so', 'because of', 'therefore', 'in order to'.

# 3  SITE AND LOCATION/SITUATION

This is seldom examined.

Referring to an OS map, you may be required to describe the site of an urban zone. If so, describe height of land, gradient, aspect and perhaps drainage (i.e. presence of stream or river and flood potential).

More likely, yet still only asked on two occasions, you may be asked, for a city you have studied, to explain the ways in which its site and location have contributed to its growth.

# EXAM EXAMPLE 5

> **For a city you have studied, explain the ways its site and situation contributed to its growth.**
>
> 8 marks

**Answer 44**   **This is an excellent answer.**

London

The earliest settlement developed around a ford across the R Thames (✔) and the dry river terraces on the north side of the river (✔) allowed the Romans to build the defensive town of Londinium (✔). Bridges were built (✔) such as London Bridge and Tower Bridge (✔), so that it became a route centre (✔) and a market town in the middle of a fertile agricultural hinterland (✔). The estuary of the Thames provided a sheltered harbour (✔), so that the town developed as a port (✔) and around the docks important industries grew up, processing imported raw materials (✔). The surrounding flat land enabled the town to expand rapidly (✔). Being located opposite the European mainland it developed important trading links (✔) and grew to be the major trading and financial centre that it is today (✔).

Total: 8 out of 8

This is an excellent answer, containing specific references to London and is well worth full marks.

You may also be asked in an OS map question for map evidence of a particular **function**, not just of a zone, but of the whole city. To date the only function examined in this way has been tourism and then only once. Other possible functions are market, port and services (e.g. university, schools, public buildings, hospitals). An industrial function would be more likely to appear in the Industry section of the exam.

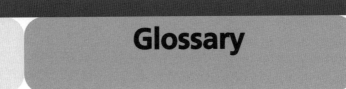

# Glossary

**Function** – What a city or urban zone 'does', i.e. what the main activity is within that city or zone.

**Hinterland** – The land surrounding a settlement.

**Location/situation** – Two terms describing the position of a settlement, i.e. where it is in relation to major features of the landscape and country.

**Site** – The land on which a settlement is built.

# 9 Rural Land Resources

*1 Landscape formation*

*Exam example 1*

*Exam example 2*

*2 Social and economic opportunities*

*Exam example 3*

*3 Environmental problems/conflicts*

*Exam example 4*

*4 National Parks*

*Exam example 5*

*5 EU/UK rural land use policies*

In Paper 2 (Environmental Interactions) there are three questions to choose from in Section 1 and only one to be answered. Question 1 on Rural Land Resources is the most popular, so that's the question which will be analysed here.

In your course you will have looked at a case study of each of the three landscape types that you studied in the Lithosphere section of Paper 1. There are usually three or four parts to this question, totalling 50 marks. The most commonly examined topics relate to **characteristics and formation of the landscape** (virtually 100% – in every paper since 1993!), the **economic and social opportunities** created by the landscape (80%) and **environmental problems/conflicts** which arise between different land uses (90%). Also, every three years there is a resource-based question on **National Parks** and, very rarely, a question on the effect of **EU and UK policies** on rural land use.

This then may seem to be very predictable. Do remember though that there are three different landscapes on which to base these questions – **glaciated uplands,** limestone uplands and coasts so there is a lot to learn. Still, the way the question on conflicts is posed, you only need to know these for two of the three landscapes – coasts and one of the uplands. Also, though you do need to know the socio-economic opportunities for all three landscapes, there is considerable similarity in the socio-economic opportunities between the two upland areas. Moreover, the landscape formations are covered in the Lithosphere section of Paper 1 and all you need to know in addition are some named examples of features in each of the landscapes. So, maybe there is not too much extra to learn!

# 1 LANDSCAPE FORMATION

Every year from 1993 onwards there has been a question relating to this. Moreover, the number of marks awarded to it has been at least 20 of the 50 available in the whole question. So the message is clear – you need to know this thoroughly, especially since this knowledge is required in the Lithosphere question in Paper 1 as well.

Coastal landscape only recently entered the syllabus and at the time of writing has yet to be examined in Paper 2, so it is a good bet that it will appear in 2007 or 2008! We shall see! Limestone has been slightly more favoured than glaciation, but it would seem likely that in future the three landscape types will be tested with equal regularity.

The question may well be posed in a rather long-winded fashion, but the gist of it will be that you have to **describe and explain with the aid of annotated diagrams the formation of the features of a particular landscape you have studied (or 'the formation of the physical landscape'**, which amounts to the same thing). There is usually a source for the question, either a quote, statement or diagram, which unfortunately is sometimes of little help to the candidate and may even be a distraction, as the first question below demonstrates. So, read the question carefully and focus on what is really required.

Before looking at any specific questions and candidate answers it is as well to understand how this type of question is marked, given a maximum of 20.

- If no annotated diagrams are used it will be marked out of 16.
- There will be a maximum of 8 marks, possibly 10 for description and explanation of one feature.
- At least three features must be covered for full marks.
- A list of features will receive a maximum of 4 marks.
- 2 marks are available for named examples of features.

- Both description and explanation are required for full marks.
- Credit might be given for features of mass wasting, such as scree formation, but there have been occasions in the past when this has not been credited, so the advice is to only include this in your answer if you have time in hand.
- In coasts and glaciation you can refer to both erosion and deposition features and in limestone to surface and underground features.

For further advice on how to describe and explain the formation of features look back at the important tips on page 41 of Chapter 3 Lithosphere.

# EXAM EXAMPLE 1

**Study Reference Diagram.**

**The Cairngorms is an area of outstanding glaciated upland scenery and has been designated as Scotland's second National Park.**

*Describe* and *explain*, with the aid of annotated diagrams, the formation of the glacial features in the Cairngorms or in any glaciated upland area in the UK which you have studied.

20 marks

*Reference Diagram (The Cairngorms mountain range)*

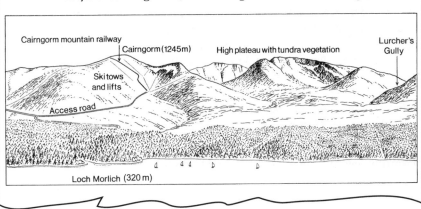

**Answer 45** **This is a weak answer.**

The Cairngorms has been designated as Scotland's second National Park because it has such spectacular scenery. The aim is that this scenery should be conserved and protected from the large number of visitors that are attracted by the mountain railway and Loch Morlich and the outdoor activities like skiing and yachting that take place here. The National Park status may encourage more people to visit the Cairngorms, boosting the local economy and providing jobs.

Glacial features that are found in the Cairngorms are corries, arêtes, pyramid peaks, U-shaped valleys, truncated spurs, hanging valleys, roches moutonnées, terminal moraines and ribbon lakes **(4)**.

A corrie is a huge armchair shaped hollow on the side of a mountain **(✔)**. An arête is a knife edged ridge **(✔)**. A pyramid peak is a pointed peak with corries on three sides **(✔)**. A U-shaped valley has steep sides and a wide fairly flat floor **(✔)**. A truncated spur is a shoulder of high land which is much steeper low down than higher up **(✔)**. A hanging valley is a tributary valley whose lip is perched high above a bigger U-shaped valley **(✔)**. A terminal moraine is a pile of loose angular boulders **(✔)**. A ribbon lake is long and narrow **(✔)**. A roche moutonnée is a rock which is steep on one side and gentle on the other **(✔)**.          Total: 10 out of 20

**Why is this a weak answer?**

This candidate has misread the question in various ways but has not been helped by the wording of the question. The instruction to study the reference diagram is of little help, since the annotations do not refer to glaciation and the glacial features that are shown in the diagram are not readily identifiable due to the scale. The references to 'outstanding scenery' and 'Scotland's second National Park' are also of no relevance to what the question is actually asking in its last sentence. So beware! There can be some fairly irrelevant stuff to plough through and you must read the question carefully to find out what is really being asked.

Unfortunately this candidate has failed to do this. In his first paragraph he has been sidetracked by the reference to the National Park and the irrelevant information in the diagram and gets no marks for this.

In paragraph 2 he is at least starting to answer the question but has only provided a list of features for which there is a maximum of 4 marks.

In paragraph 3 he goes on to describe the appearance of these features, but there is no explanation of their formation, so he can get no more than half marks. He has also failed to give any named examples and has not used any annotated diagrams. In short, he is probably fortunate to receive 10.

Look at Chapter 3 Lithosphere to see an example of how to use annotated diagrams to explain the formation of a corrie.

# EXAM EXAMPLE 2

> **Several National Parks are particularly noted for their glaciated or upland Limestone scenery.**
>
> **Choose one of these types of landscape and, with the aid of *annotated* diagrams, explain how the main features of the physical landscape were formed.**
> 20 marks

**Answer 46**   This is a good answer.

The Yorkshire Dales are an upland limestone landscape. Limestone consists of calcium carbonate (✔) which reacts with rainwater, a dilute form of carbonic acid (✔) so that the rock is dissolved (✔). Limestone is a sedimentary rock that was deposited in layers separated by bedding planes (✔). During the Ice Age glaciers scraped away the overlying soil in some places exposing the top bedding plane to create a pavement (✔) like that found above Malham Cove (✔). There are also vertical weaknesses in the rock beds called joints (✔), into which rainwater drains and dissolves the limestone to widen the joints and form grooves in the pavement called grikes (✔). The undissolved blocks between the grikes are called clints (✔). Where the glaciers have deposited boulder clay (✔) in places it has slipped down into solution hollows forming shakeholes (✔). There is very little surface drainage (✔) because rainfall drains underground through the permeable limestone (✔). However, towards the end of the Ice Age the ground was still frozen and impermeable (✔) so that meltwater rivers flowed over the surface and eroded V-shaped valleys (✔). Once the climate warmed up the water was able to drain underground again, leaving dry valleys (✔) like Watlowes in the Malham area (✔). The point at which a stream disappears underground is called a swallow-hole (✔), e.g. Gaping Ghyll (✔). Gorges, such as that at Gordale Scar (✔), form where underground tunnels have collapsed (✔).        Total: 16 out of 20

**Why is this a good answer?**

The only failing of this answer is that the candidate has not provided annotated diagrams, which is a costly oversight, since she has lost 4 marks as a result (see page 44 for an example of using an annotated diagram to explain the formation of a stalactite). Nevertheless, this is a lucid well-written answer which describes and explains the formation of several landforms, a couple, the pavement and dry valley, in some detail. She has also given four named examples and receives the 2 marks available for these. Do not think that you only need to give 2 examples – we do not know for certain that 2 marks is all that will be awarded in future questions and it is always good practice to give examples where you can, because it creates a good impression on the examiner.

Also, you should note that she has only referred to surface features and perhaps forgotten to mention underground features, such as caves, caverns, potholes, stalactites and stalagmites. Other features not mentioned are scars and screes. So, you see, if you have revised thoroughly, you could get carried away in a question like this and spend a disproportionate amount of time writing far more than is needed – watch the clock!

A sensible approach to answering this question is:
- use annotated diagrams;
- describe and explain in detail the formation of three features (perhaps four if you have time);
- quickly mention any other features that you know;
- give named examples of four or five features.

# 2 SOCIAL AND ECONOMIC OPPORTUNITIES

This topic also appears very frequently and is usually worth 10 or 12 marks, so be prepared.

According to the syllabus you need to know social and economic opportunities for a case study of each of the three landscape types. However, for a number of years now, the wording of the question in the exam has only required you to know them for a 'protected upland area which you have studied' (i.e. a National Park in either a glaciated upland such as Snowdonia or a limestone upland such as the Yorkshire Dales). Consequently you can probably get away with only revising this topic for the new coastal case study and **either** the glaciated case study or the limestone case study. Even if you get caught out you should still be able to answer well, because there is a strong similarity between the opportunities in the glaciated and limestone uplands.

There are three ways in which this topic is examined.

i)  Most commonly you are simply asked to 'explain the social and economic opportunities created by the landscape'. Make sure you refer to both social (e.g. recreation, nature reserves, military training, tourism) and economic (e.g. farming, forestry, fishing, water supply, HEP, mining, quarrying). For each land use you should explain how the landscape encourages it. Thus, you will tend to stress positive features, e.g. 'The spectacular scenery of arêtes, pyramid peaks and ribbon lakes attracts both tourists and climbers to the area'. Negative features may also be appropriate, e.g. 'The thin soils and cold wet climate are not suited to arable farming, but sheep can feed on the rough grazing of the hills'.

> In this question weak answers often only refer to tourism/recreation and forget that there are other social and economic opportunities as shown above.

ii) You are asked to explain 'how environmental factors limit the social and economic opportunities' or similarly 'how physical factors limit human activity' or 'how the physical landscape restricts development'. These different wordings all mean the same thing. In answering you will tend to stress the negative features, e.g. 'The high mountains and winter snowfall make road transport difficult', 'The high altitude causes low temperatures and a short growing season, which, combined with thin acid soils and lack of flat land makes arable farming difficult', 'Steep slopes mean that farmers cannot use machinery, limiting the possibilities for growing crops', 'Forestry cannot be done above 500 m because higher land is exposed to the wind', 'The remote location hinders industrial development because of high transport costs'.

iii) Occasionally the question just focuses on tourism and asks you to explain how it benefits an area. You should refer to such things as improvements in the economy through expanding business for shops, hotels, restaurants, B and Bs, camp sites, etc; new jobs in catering, guided tours, information centres, etc; and improvements to the infrastructure such as new roads, by-passes, better bus services, public toilets, leisure centres, etc; and the possibility that rural depopulation will be halted.

# EXAM EXAMPLE 3

> **For a coastal area that you have studied, explain the social and economic opportunities created by the landscape.**
>
> 10 marks

**Answer 47**   This is a weak answer.

On the south coast of England there is nice scenery which attracts tourists. They bring money into the area and provide jobs for local people such as waiters, tour guides and bar staff. Hotels and guest houses will get a lot of trade and shops can sell souvenirs. This helps the economy. Visitors will eat in the local restaurants so they will make more money too. With more money services like roads will get better.

Total: 2 out of 10

**Why is this a weak answer?**

This is a response typical of a weak candidate. It is vague and does not properly answer the question. It would be more acceptable if the question were 'describe the benefits of tourism', but that is not the question. The answer must refer to the landscape. Only the first sentence of this answer does this and then only vaguely – what type of scenery? No named examples have been given and even the coastal area chosen is somewhat imprecise. Some credit would be given for developing the idea of tourism opportunity, but not much, since the answer should focus on the landscape and how this influences the land use. Consequently it only receives 2 marks.

**Answer 48**   This is a much better answer.

Tourism and recreation are the main activities on the Dorset coast. It has attractive and varied scenery such as the arch at Durdle Door (✔) and the stacks at Old Harry's Rocks (✔). These are appreciated by walkers who come to use the South West Coast Path (✔). Lagoons like West Fleet (✔) are nature reserves, good for birdwatching (✔). Other recreational activities are rock climbing on the limestone sea cliffs (✔) and swimming in the sheltered water of Lulworth Cove (✔). Poole Harbour is also sheltered (✔) and an excellent location for a marina used as a base for sailing yachts (✔). Boat cruises also run from here catering for the tourists (✔). Swanage has a long beach (✔) ideal for families, so the town has developed as a holiday resort with many hotels, restaurants and amusement arcades (✔), which provide many jobs for local people as bar staff, waiters etc and boosts the local economy (✔). Limestone is quarried on a large scale at Portland (✔), which is a good location since it can be shipped out easily by sea (✔). The sedimentary rocks also contain oil which is extracted at Wytch Farm (✔). The gentle hills of the South Downs which fringe the coast are ideal for military training (✔) and sheep can be farmed on their thin soils (✔).               Total: 10 out of 10

**Why is this a much better answer?**

This excellent answer has actually produced more than is needed and is well worth full marks. Notice how each activity and land use is linked to a landscape feature, with plenty of named examples of features and places. Both social and economic opportunities are covered.

# 3 ENVIRONMENTAL PROBLEMS/CONFLICTS

In only one year has there not been a question which relates in some way to this topic and there are often 20 marks for it, so, like the previous two topics, it is worth revising thoroughly.

You need to have knowledge of two case studies – one of a **coastal** area and one of an **upland** area.

Most commonly you are asked to **explain the conflicts** between different land uses, **describe solutions/measures to resolve these problems** and occasionally comment on how effective they have been. Examples of relevant land uses are tourism, recreation, forestry, farming, military training, mining and quarrying. In recent years, however, the questions have tended to focus on the environmental problems and conflicts created by **tourism**, recreation and large visitor numbers, so perhaps you should prioritise this in your revision. Knowledge of the role of National Parks and their aims can also be relevant here.

# EXAM EXAMPLE 4

> **For any *named* upland area of the UK which you have studied,**
>
> i) *give examples* of environmental conflicts which have arisen, and
>
> ii) *describe* some of the measures taken to resolve these conflicts and *comment* on their effectiveness.     20 marks

**Answer 49**   **This is a good answer.**

The Lake District attracts large numbers of visitors who can cause considerable traffic congestion on the narrow roads (✔), e.g. in Langdale (✔). The noise of the traffic disturbs the peace (✔) in villages like Ambleside (✔). To overcome this problem in Keswick they have built a by-pass to the north of the town (✔). This has reduced the congestion, but south bound traffic still has to go through the town (✔). The creation of a one way system and a new car park has also helped and allowed visitors better access to the shops (✔). The increasing numbers of hillwalkers have eroded paths (✔) on mountains like Helvellyn and Scafell (✔), but this is being tackled by repairing paths with terram and local stone (✔), which blends nicely with the landscape (✔). Motor boats on the lakes also disturb the peace and their waves erode the banks (✔). Consequently the National Park only allows motor boats now on Lake Windermere (✔). Providing plenty of facilities like toilets, car parks, picnic areas, litter bins etc (✔) in specific honeypot areas like Tarn Howes (✔) concentrates visitors in a few areas and allows other areas to remain peaceful and the wildlife undisturbed (✔).

Commercial forestry in Ennerdale has spoiled the natural beauty of the valley (✔) by blanket planting of angular blocks of coniferous trees (✔). This has removed the moorland habitat of rare birds like merlin (✔). New planting has irregular edges, follows the contours of the land, leaves ridges unplanted and puts deciduous trees at the edges (2). This has the effect of making the landscape look more natural (✔). Slate is quarried at Honister (✔) which scars the landscape and is noisy (✔), so the times for blasting are restricted (✔).        Total: 20 out of 20

**Why is this a good answer?**

This answer covers all aspects of the question. The candidate has given plenty of named examples and gets the 2 marks for these and has managed to comment sufficiently on the effectiveness of the solutions. He has run parts (i) and (ii) together in his answer, describing each conflict and then immediately suggesting solutions and commenting on their effectiveness. This means that he does not have to repeat himself and it is perfectly acceptable since there is a holistic mark for parts (i) and (ii). If, however, each part were given a separate mark, then you would have to answer each part separately.

If the question had been restricted just to conflicts created by tourism he could have added to what is in paragraph 1 by referring to such things as air pollution from car exhausts, problems to farmers like sheep disturbance in the lambing season and wall damage, litter, new development eyesores, sale of holiday homes pricing locals out of the market and trampling of flowers, and other solutions like educational visitor centres, wardens and park and ride schemes.

# 4 NATIONAL PARKS

You should be familiar with the aims of National Parks, their distribution and reasons for location, the arguments for and against their designation and the way they can resolve environmental conflicts. It is rare for a question to be asked directly about this, although it did occur in 2004. Much more common (every three years on average) is a resource-based question with a map or table. Before 1998 these tended to be on land use or land ownership patterns in different National Parks, but since then the focus has been on the contrasts in visitor numbers/popularity between the National Parks.

There are usually 10 marks available, which, on first examination of the resources, can seem difficult to achieve. Just remember the Enquiry Skills you developed for the Standard Grade exam and take as much as you can from the resource. Look at all the data provided and consider how it could be used in your answer.

# EXAM EXAMPLE 5

Study Reference Map and Reference Table.

*Explain* why the number of visitors to National Parks can vary so greatly. 10 marks

Reference Map (National Parks in England and Wales)

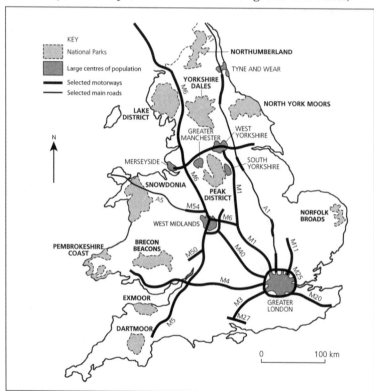

Reference Table (General National Park Statistics)

|  | Brecon Beacons | Dartmoor | Exmoor | Lake District | Northumberland | North York Moors | Peak District | Pembrokeshire Coast | Snowdonia | Yorkshire Dales | Norfolk Broads |
|---|---|---|---|---|---|---|---|---|---|---|---|
| Designation Year | 1957 | 1951 | 1954 | 1951 | 1956 | 1952 | 1951 | 1952 | 1951 | 1954 | 1989 |
| Area (ha) | 135144 | 95570 | 69280 | 229198 | 104947 | 143603 | 143833 | 62000 | 214159 | 176869 | 30292 |
| Visitor days (millions per year) | 7 | 8 | 3 | 20 | 1.5 | 11 | 22 | 13 | 8 | 9 | 5 |

**Answer 50** This is a very good answer.

The most popular parks are the Peak District and Lake District with 22 and 20 million visitors a year (✔). Both were established in 1951 when the first parks were set up and have had a long time to develop their facilities, unlike the less popular Norfolk Broads which was not established until 38 years later (✔). They are the biggest and fourth biggest parks so have a larger number of places within them for tourists to visit (✔). Many people can easily visit the Peak District since it is surrounded by large conurbations like Greater Manchester and S Yorkshire (✔) and is one of the closest parks to London (✔), being only 200 km away (✔). Consequently many people can visit just for the day (✔). The Lake District has many different attractions, appealing to a wide range of interests, like boating on the lakes, rock climbing and hillwalking and visiting picturesque villages like Grasmere (✔). Also the glaciated scenery is more dramatic than most other parks (✔), with ribbon lakes, pyramid peaks and arêtes (✔). The Peak District is the most accessible park, having motorways on all sides (✔) and the Lake District is close to the M6 (✔). Some of the least accessible parks (e.g. Norfolk Broads, Exmoor and Northumberland) have lower visitor numbers (✔). The third most popular park is the Pembroke Coast, which has spectacular cliff and arch scenery (✔) and big safe beaches for families like Pendine Sands (✔).

Total: 10 out of 10

## Why is this a very good answer?

The candidate has used the data very well. From the map she has seen that motorways are marked and concluded that accessibility is significant and also noted that large centres of population are shown, suggesting that proximity of these to the parks is also important. She has used information from the table, identifying the most and least popular parks and made sensible statements relating to their areas and even their designation years. She has also worked the map by using the scale to estimate distance and has quoted some figures from the table for which there may be some marks available. She has also managed to slip in some of her case study knowledge with her references to scenery in the Lake District and Pembroke Coast.

# 5  EU/UK RURAL LAND USE POLICIES

This topic has only been examined on two occasions, the last time in 2000, so it may be appearing again soon! It will probably be worth 10 or 12 marks. Set Aside and Quotas are the schemes most likely to be tested, but other possibilities are Farm Diversification Grants, Woodland Grant Schemes, Environmentally Sensitive Areas and Less Favoured Areas. Revise a minimum of three of these. You should be able to state the reasons for and the aims of the policy and be able to discuss its good and bad points.

# 10 Development and Health

**1 Differences in levels of development between and within ELDCs**

   *Exam example 1*

**2 Development indicators**

   *Exam example 2*

**3 Disease**

   *Exam example 3*

**4 Health care**

   *Exam example 4*

**Glossary**

Section 2 of Paper 2 also has three questions to choose from. Question 6 on Development and Health is by far the most frequently answered question, so is the question analysed here.

There are four main topics covered and only three or four parts to the question, totalling 50 marks, making it fairly predictable.

## 1 DIFFERENCES IN LEVELS OF DEVELOPMENT BETWEEN AND WITHIN ELDCS

In the last 13 years there has been only one year when this topic has not been tested, so revise it thoroughly. On a couple of occasions the question has been about differences within **one** ELDC that you have studied, but more commonly (10 times in 13 years) it has concerned differences **between** different ELDCs. Thus you need to know:

● a case study of one ELDC with explanations of the differences in development within that country;

- specific examples of better developed ELDCs and the reasons for their relative wealth;
- specific examples of poorer ELDCs and the reasons for their relative poverty.

**Read this question carefully**. Many candidates will write about differences between countries when they should be explaining differences within one country. **So look out for those key words <u>between</u> and <u>within</u>**.

Often there is a Reference Table of Development Indicators for a number of countries. Again, read the question carefully, because it may be that you can refer to any ELDCs you have studied – not just those in the table.

# EXAM EXAMPLE 1

**Study Reference Table.**

**The Reference Table shows data for three countries which, although ELDCs (Economically Less Developed Countries), are relatively rich.**

**Giving examples from named countries, *describe* the factors which help some ELDCs to achieve higher levels of development than others.**                                      10 marks

*Reference Table (Development indicators for selected countries)*

| Country | Birth Rate (per 1000) | Infant Mortality Rate (per 1000 live births) | Literacy Rate (%) | GNP per capita ($US) |
|---------|-----------------------|-----------------------------------------------|-------------------|----------------------|
| Thailand | 16 | 21 | 93 | 7400 |
| Saudi Arabia | 30 | 14 | 79 | 11 800 |
| South Korea | 12 | 7 | 98 | 17 700 |

**Answer 51**    **This is a weak answer.**

S Korea is the wealthiest country, having a GNP of $17700 per capita. Consequently it only has a Birth Rate of 12 per 1000 because it can afford hospitals and the Infant Mortality Rate is also the lowest. It also has the highest literacy rate because it has invested more money in health care and education than the others (✔). Thailand has the lowest GNP, being poor and so has the highest Infant Mortality Rate. Saudi Arabia has oil supplies so has a higher GNP than Thailand (✔).    Total: 2 out of 10

#### Why is this a weak answer?

This candidate has not read the question carefully enough. We are told that all the countries in the table are relatively rich ELDCs. The question is to 'describe the factors' that help richer ELDCs become more developed than the poorer ones, but this answer mainly just identifies connections between different data in the table (which might be appropriate in other questions but not this one). Only in the references to investment in health and education and Saudi Arabia's oil supplies are 'factors' described.

It also instructs us to give examples from named countries – it does not stipulate that you should only refer to those in the table, which is what this candidate has done. If your reaction on reading a question is 'Help! I've not studied any of these countries', look again at the question and you will certainly find that you can refer to any countries that you have studied. There are a multitude of ELDCs and the examiners cannot expect you to know about all of them, but they do expect you to have studied some of them.

**Answer 52**    **This is a very good answer.**

The amount of natural resources that a country possesses will influence its ability to develop (✔). Saudi Arabia and Kuwait have large oil reserves to export (✔). Oil is in great demand so they can sell this for a lot of money, giving them a high GDP (✔), but it only goes into the pockets of a few people in the country and most people stay poor.

Wars prevent a country from developing so Iraq is not getting the benefit of its oil reserves (✔) and Somalia has been held back by its civil war (✔).

Some of the poorest countries in the world are in the Sahel zone of Africa like Mali and Chad (✔). These countries have severe climatic problems with frequent droughts (✔). These mean they cannot produce enough food for their populations and have to borrow money for this instead of investing the money in development projects (✔).

China is an NIC developing its industry rapidly by allowing multi-national companies to invest there (✔), but this is having a damaging effect on the environment with serious pollution of the air and rivers causing ill health in the industrial areas. The Yangtse River dolphin is almost extinct.

Singapore has the advantage of a natural harbour in a strategic position (✔), enabling it to develop its port with trading links to many other countries (✔). Cheap labour and imported raw materials (✔) enabled it to develop its textile and shipbuilding industries (✔). It is a 'tiger economy' (✔) which has used its entrepreneurial skills to develop as a commercial centre (✔).

Total: 10 out of 10

## Why is this a very good answer?

In paragraphs 1 and 4 there are irrelevant references to uneven distribution of wealth and pollution, not to mention the poor old Yangtse River dolphin.

Otherwise this is a generally sound answer which identifies factors and backs them up with examples. The last paragraph is particularly impressive with its more detailed knowledge of Singapore. Had no examples been given it would have been marked out of 8.

The candidate has also realised that he does not necessarily have to refer only to countries in the table.

Other factors that could have been mentioned are the frequency of natural disasters, strategic plans such as India's 5 year plan, a successful Green Revolution in some countries and the fact that some countries benefit more from aid than others.

# 2  DEVELOPMENT INDICATORS

This topic is also tested frequently, in two out of every three years on average. Make sure you know the difference between **social** and **economic** indicators (see Glossary).

You may be asked to:

- Identify indicators which could be used to produce a **composite index** such as the HDI or PQLI. This is easy but you need to be precise, e.g. 'GDP per capita' not just 'GDP', 'Adult Literacy Rate' not just 'Literacy'.

- Explain how the indicators illustrate a **country's level of development** or say how useful they are, e.g. 'Percent of population in Primary employment – in ELDCs many people are engaged in subsistence agriculture so the figure will be high, whereas in EMDCs the figure will be low because most work on farms is done by machines and fewer agricultural workers are needed.'

- Explain why certain indicators 'may not give an accurate representation of the **true quality of life** in an area' (or 'do not **identify different levels of development** within a country', which is the same thing). This is the question which has been most commonly asked. The number of marks has come down over the years but might be as many as 10.

# EXAM EXAMPLE 2

> **Explain why indicators of development may fail to provide an accurate representation of the true quality of life <u>within</u> a country.**
> 10 marks

**Answer 53**    **This is a weak answer.**

You cannot trust the figures in many ELDCs because the census data may be inaccurate (✔). This is because a census is difficult to carry out in poor countries due to the expense (✔) and they can only afford to do it once in a while and then, as in Nigeria, it may only be a sample. Consequently the information is often out of date. Some areas of a country may be inaccessible, perhaps mountainous with few roads. Literacy is low in ELDCs so many people may not be able to fill in forms. Also there are many nomadic people in N African countries so they may be missed by the people carrying out the census. Civil war can also prevent a census being carried out.                        Total: 2 out of 8

**Why is this a weak answer?**

It is certainly valid to point out the significance of inaccurate census results, but the examiners will only award a maximum of 2 marks for problems of collecting census information. This candidate has become fixated with these difficulties and has lost sight of the main thrust of the question.

**Answer 54**    **This is a better answer.**

The development indicators are averages (✔). Therefore the figures may be much higher than the average in one part of the country and much lower in another (✔). For example in Ethiopia the capital city Addis Ababa has a lower death rate than the average because of better access to hospitals and doctors (✔). In the North East of the country it is higher than average since there are fewer medical facilities and there is a civil war going on (✔). In Italy there is a N-S divide between the wealthy industrialised north and the poorer more rural Mezzogiorno (✔). The capital city of an ELDC often has more favourable statistics than the rest of the country because the government is based there (✔) and will be more likely to provide facilities in the capital before it can afford to do so elsewhere (✔). However, even within cities, there can be big differences e.g. in Lima there is much lower life expectancy in the 'pueblos jovenes' slums like Cono Norte than there is in the wealthy area of Miraflores (✔). In Saudi Arabia vast revenues from oil gives a high GNP per capita (✔), but this money only goes to a few extremely rich families, leaving the rest of the country in poverty (✔).        Total: 8 out of 8

**Why is this a better answer?**

This is a good answer, the use of examples being superior to bland references to N–S divides, urban–rural contrasts and rich minorities. The answer, though, could also have mentioned that certain indicators may not actually be relevant to the real quality of a person's life.

# 3 DISEASE

You need detailed knowledge of **one** water related disease – either **malaria**, **cholera** or **bilharzia/schistosomiasis**.

In most years (80%) there is a specific question on the disease and there may be as many as 30 marks available! Need you be told to revise this topic thoroughly?

You need to know:

- The physical/environmental and human factors which have 'put people at risk of contracting the disease' (or 'caused the spread of the disease', which is virtually the same thing).
- The measures/strategies used to control/combat the disease.

- How effective these measures have been (or an 'evaluation of strategies' – the same thing).
- The benefits to ELDCs of eradicating the disease. This is much less frequently asked, but straightforward to answer along the lines of a healthier population leading to greater productivity, an improved economy and a better standard of living.

# EXAM EXAMPLE 3

---

For *either* malaria *or* bilharzia (schistosomiasis) *or* cholera,

i) *describe* the environmental *and* human factors which put people at risk of contracting the disease,

ii) *describe* the methods used to try to control the spread of the disease, and

iii) *comment* on how effective these methods have been.

24 marks

---

**Answer 55**   **This is a weak answer.**

i)  Mosquitoes pass on the disease and they breed in water in hot climates (✔) where there is shade (✔).

ii)  Spraying, drainage, plant trees, mustard, coconuts, drugs, nets **(2)**.

iii)  Less people killed now than before (✔).   Total: 5 out of 24

**Why is this a weak answer?**

This is a candidate who has badly misjudged her time and spent far too long answering the other parts of the paper and, it would appear, left only a few minutes for this hastily scribbled response. She has not even said which disease she has chosen. There is no detail in any part of the answer and her list of methods is so vague that it only warrants 2 marks. They do hint at a greater depth of knowledge, so it is a shame that the candidate did not leave herself time to demonstrate it.

**Answer 56**    **This is a better answer.**

Malaria

(i)  The disease is passed on by the female Anopheles mosquito (✔) which needs temperatures between 15C and 40C (✔) and stagnant water for breeding (✔) such as padi fields (✔).

(ii)  Choloroquine is a drug given to sufferers (✔) to kill off the parasites in their blood (✔). People are encouraged to sleep under mosquito nets at night (✔) to prevent being bitten when the mosquitoes tend to feed most often (✔). The breeding areas were sprayed with insecticides like DDT (✔) and swampy areas have been drained to reduce the number of places they could breed (✔). Another way to remove water is to plant trees which will soak it up through their roots (✔). The eucalyptus tree is particularly effective (✔). Another solution is to infect coconuts with bti bacteria (✔), split them and throw them into the water so that the mosquito larvae eat the bacteria and are killed (✔). In India they put small fish into the padi fields to eat the larvae (✔), or sprinkle mustard seed on the surface (✔), which sinks and drags the larvae down, drowning them (✔). Yet another way to kill them is to suffocate them by spraying egg-white on the surface of the water (✔).

(iii)  Mosquitoes are becoming resistant to insecticides (✔) so that in countries like Sri Lanka malaria is increasing again (✔) and there are still over 400 million people in the world suffering from the disease (✔). DDT was very successful at first (✔) but did a lot of damage to wildlife and the environment by getting into the food chain and is now banned (✔). The coconut method has the advantage of being cheap (✔).

Total: 24 out of 24

**Why is this a better answer?**

There would be up to 14 marks available for one part of the question and all three parts would have to be answered for full marks. Part (i) is rather weak, failing to identify the climatic areas where the disease occurs, the mosquito's need for shade and other man-made breeding areas such as irrigation channels, reservoirs, and bomb craters. However, part (ii) is excellent, getting the full 14 marks available so that the whole answer just manages to gain full marks.

# 4 HEALTH CARE

Roughly every two years there is a question relating to **Primary Health Care**.

You need to be able to:

- describe and explain the main strategies/methods used in Primary Health Care, showing particularly why they are appropriate in ELDCs;
- give examples of places where they are used;
- comment on their effectiveness.

In addition, relating to the general conditions which lead to health problems, you need to know:

- how lack of clean water and inadequate sanitation leads to high levels of disease;
- the physical and human factors which lead to ill health, malnutrition, early death, high childhood mortality rates and variations in life expectancy in ELDCs.

# EXAM EXAMPLE 4

**Study the Reference Diagram which highlights the main aspects of Primary Health Care (PHC).**

*Give examples* **of Primary Health Care strategies and suggest why such approaches to improving health standards are suited to** *Developing Countries.* **12 marks**

*Reference Diagram (Elements of Primary Health Care)*

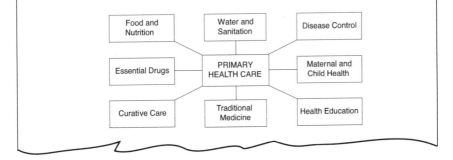

**Answer 57**   **This is a weak answer.**

Primary Health Care strategies are suitable approaches to improving health standards in ELDCs. Giving essential drugs can cure diseases. Food and nutrition is important because if people have good food they will be healthy. Water and sanitation is needed for clean living. If people are educated they will know what to do.

Traditional medicines are used which are cheaper than modern drugs (✔) which is good in ELDCs since they do not have the money to pay for these (✔).

Other things which are done are Maternal and Child Health and Curative Care.

Total: 3 out of 12

**Why is this a weak answer?**

The reference diagram is a helpful prompt suggesting topics which the candidate can enlarge on, but this candidate seems unable to do this. Paragraph 1 starts by repeating the question and then makes a series of extremely vague statements, none of which would be worth a mark on their own, but which might just gain an overall token mark. Paragraph 3 merely lifts information straight from the diagram. Only in paragraph 2 does she make any valid points and then only simply, without exemplification.

**Answer 58**   **This is a better answer.**

Primary Health Care was pioneered in China in the 1960s because people living in the countryside had no access to health care (✔). The few hospitals were located in the larger towns and cities like Chengdu, which could be many days travel away (✔) and anyway there were not enough trained doctors to treat all the people (✔). Therefore they trained local people in basic hygiene and first aid (✔). They were known as 'barefoot doctors' and they could treat common ailments (✔) and man a network of small local clinics so that a very basic form of health care could be provided (✔). They could also use traditional medicines using local ingredients which were much cheaper than modern drugs (✔). It costs much less to train these people than it does to train a proper doctor (✔) so it is the only economic way for a poor country to provide health care for the majority of its population (✔). It also means that hospitals and trained doctors can concentrate their resources on more complex cases so is more efficient (✔).

Many diseases in ELDCs such as cholera and diarrhoea are caused by contaminated water (✔). Therefore the Indian government is encouraging self-help schemes (✔) to sink tube wells to reach pure underground water (✔) and build village toilets with septic tanks to keep drinking water and sewage separate (✔). Babies can be treated by ORT, a solution of water, sugar and salt (✔) that can be cheaply made locally and will cure diarrhoea (✔).

Total: 12 out of 12

## Why is this a better answer?

This is a good answer. Had it been needed the candidate could also have referred to vaccination programmes against diseases such as polio, measles, and TB and the importance of health education.

Although the vast majority of questions are on the four specific topics now detailed, there are occasionally differently slanted questions which require you to draw on more general knowledge of conditions in ELDCs. An example of this was in 2002, when the question asked the candidates to describe and suggest reasons for the differences in the provision of social services between urban and rural areas. On another occasion candidates had to explain the factors which had led to malnutrition in ELDCs and the resulting downward spiral into sickness and poverty.

# Glossary

**Economic Indicators** – These relate to money and employment, e.g. Gross Domestic Product per capita, Percentage of working population employed in the Tertiary Sector.

**ELDC** – Economically Less Developed Country, referred to in exams before 2005 as a Developing Country.

**EMDC** – Economically More Developed Country, referred to in exams before 2005 as a Developed Country.

**Gross Domestic Product (GDP) or Gross National Product (GNP)** – The value of all goods and services produced in a country.

**Human Development Index (HDI)** – A composite index based on adult literacy rate, average life expectancy and average income adjusted to spending power.

**Human factors** – Factors relating to people such as cultural habits, living conditions, population movements and construction (e.g. dams creating new breeding areas for mosquitoes).

**Newly Industrialised Country (NIC)** – A country in which the GDP from industry exceeds one third of the total GDP.

**Oral Rehydration Therapy (ORT)** – A cheap way to treat dehydration and diarrhoea with a solution of water, salt and sugar.

**Physical/environmental factors** – Natural factors such as climate, landscape, drainage.

**Physical Quality of Life Index (PQLI)** – Another composite measure of development based on Adult Literacy Rates, Average Life Expectancy and Infant Mortality Rate.

**Social Indicators** – These relate to the general health and welfare of people, e.g. Crude Birth Rate, Infant Mortality Rate, Number of TV sets per 1000 people.

# Conclusion

You should now be well placed to fulfil your potential in the external examination. If successful, you will have demonstrated that you can work well under the pressure of time, analysing data and producing reasoned responses to sophisticated questions. What is more, you will have done this in a subject which bridges sciences and the arts, drawing on a wide range of skills and giving you a sound understanding of the world around you. Not only is this a qualification valued by universities and employers alike, but it makes you a well-informed member of global society.

**Now, good luck in your exam and watch that clock!**

TIME FLIES!